Graham Steell

The Physical Signs of Cardiac Disease

Second Edition

Graham Steell

The Physical Signs of Cardiac Disease
Second Edition

ISBN/EAN: 9783337270841

Printed in Europe, USA, Canada, Australia, Japan

Cover: Foto ©berggeist007 / pixelio.de

More available books at **www.hansebooks.com**

THE

PHYSICAL SIGNS OF CARDIAC DISEASE

THE

PHYSICAL SIGNS

OF

CARDIAC DISEASE

BY

GRAHAM STEELL, M.D. Edin.

FELLOW OF THE ROYAL COLLEGE OF PHYSICIANS OF LONDON ; PHYSICIAN TO THE
MANCHESTER ROYAL INFIRMARY ; AND LECTURER IN CLINICAL
MEDICINE, OWENS COLLEGE.

FOR THE USE OF CLINICAL STUDENTS

Second Edition

MANCHESTER : J. E. CORNISH
1891

TO THE

STUDENTS OF THE MANCHESTER ROYAL INFIRMARY

FOR WHOM IT WAS WRITTEN

I DEDICATE

THIS LITTLE WORK.

CONTENTS.

ANATOMICAL CONSIDERATIONS.

THERE are certain essential points in the anatomy of the heart which it is important for the clinical student to bear in mind. What may be termed "applied" anatomy will alone be considered here.

As to the Position of the Heart and Great Arteries in the Chest.
—The heart may be regarded as an irregular cone in shape, and its longitudinal axis from base to apex runs from above downwards from right to left, and from behind forwards, in the cavity of the chest. The base is on a level with the fourth, fifth, sixth, seventh, and eighth dorsal vertebræ. The apex points in the direction indicated, and is practically in contact with the chest-wall (a small portion of the border of the left lung intervening over the very extremity) in the fifth intercostal space, about two inches to the left of the left border of the sternum. Much the larger portion of the heart is made up of the ventricles, the right being anterior, the left posterior and coming to the front only at the left border and at the apex. When the heart is regarded from the front, *in situ*, the left auricle is concealed from view, with the exception of its appendix, which appears above the left or superior cardiac border to the left of the pulmonary artery. The right auricle, covered by lung, is close to the anterior wall of the chest, and its position is of considerable importance to the physician, as will be afterwards shown. Taking the heart as a whole, two

B

thirds lie to the left of the median line of the body ; one third
lies to the right.

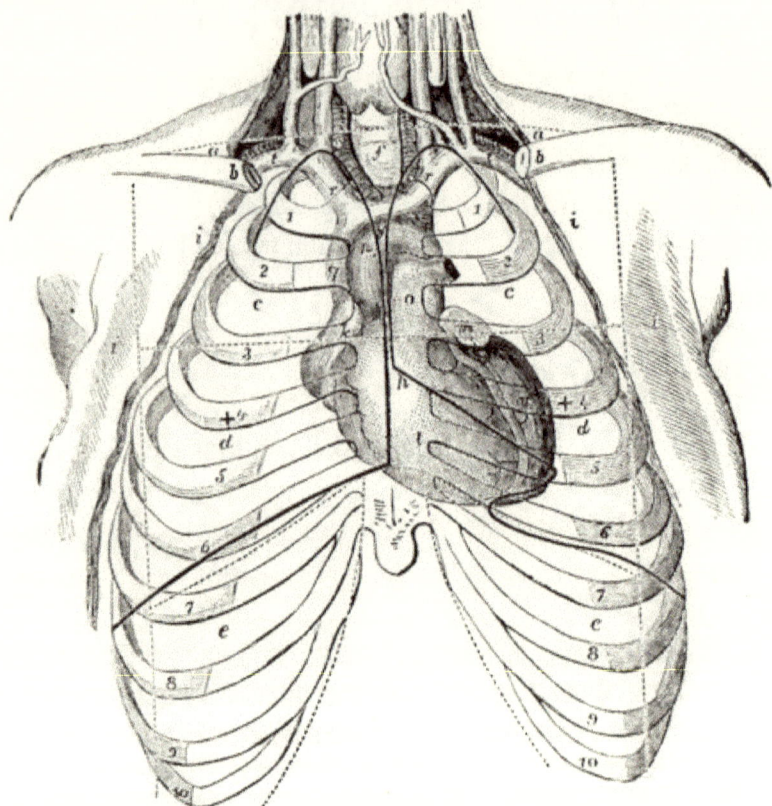

Fig. I. (from Walshe) shows the general position of the heart in the thorax. The area
of "superficial cardiac dulness," *i.e.*, the portion of cardiac surface uncovered by
lung, is also represented.

We have described the heart as a cone, and the heart proper
is of this form ; but, superimposed upon it, we have the smaller
globular mass formed by the great arteries (*vide* Fig. I.)

A glance at Figure II. will show the very oblique direction
of the tricuspid orifice. In accordance with this direction of the
orifice the blood current from the right auricle into the right
ventricle is almost horizontal, while the current of blood from

the ventricle into the pulmonary artery passes upwards, and to
the left nearly at a right angle with it. It is possible that the
latter fact bears some relation to the production of the epigastric impulse to be afterwards described. It will be evident how
enlargement of the right auricle from over-distension becomes
appreciable by percussion as an increase of the cardiac dulness
to the right of the sternum. Of all the cavities of the heart, it
is most liable to temporary over-distension, and its enlargement
is an index of the degree of obstruction suffered by the circulation in cardiac and pulmonary diseases. The thin walls of
the auricles are in marked contrast with the thick walls of the
ventricles.

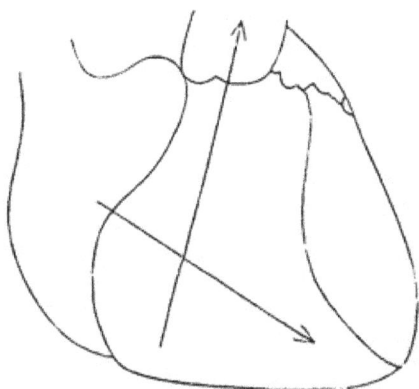

Fig. II. (modified from Sibson) is a diagrammatic representation of the circulation in
the right side of the heart.

Let us now examine a transverse section of the heart through
the auricles, and note the arrangement of the four orifices
(*vide* Fig. III.). The separation of the pulmonary valves from
the corresponding auriculo-ventricular set, is somewhat surprising
at first sight ; but a moment's consideration will tell us that this
is due to the conus arteriosus (infundibulum) of the right ventricle, which passes in front of the aortic orifice. We shall find

later that aortic regurgitation murmurs are very well heard over
the sternum, especially in its middle and lowest thirds ; and the
relation of parts which we have just described probably explains
the phenomenon. During the diastole, the pulmonary valves
being competent, there is no current of blood in the infundibulum, and it is the portion of the right ventricle in
closest relation with the sternum : an aortic regurgitation
murmur will thus be readily transmitted to the surface
through the infundibulum, and once having reached the sternum
it will be readily conducted along the bone.

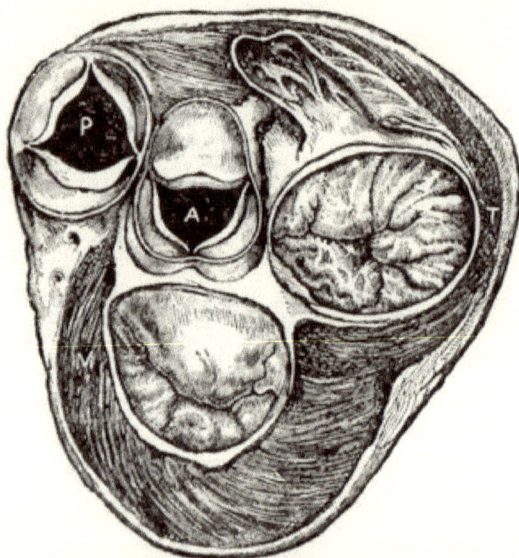

Fig. 111. (from Heath's " Practical Anatomy") shows the relative position of the
different cardiac orifices. It will be evident why aortic regurgitation murmurs are so
well heard over the sternum, as they have only to pass through the infundibulum
of the right ventricle, where, during the diastole, there is no movement of blood.
Having once reached the sternum, such murmurs are well carried along the bone.

The arrangement of the cusps of the arterial valves—two
anterior and one posterior in the pulmonary artery, two
posterior and one anterior in the aorta—is worthy of note.
The orifices of the coronary arteries open from behind the

anterior and left posterior aortic cusps. Immediately behind the two posterior cusps of the aorta is situated the large anterior flap of the mitral valve, which is embraced by the short semilunar posterior flap of the same. The close proximity of the two sets of valves should be borne in mind.

It is to be noticed that the currents of blood to and from the left ventricle are almost parallel in the long axis of that cavity, thus differing from the like currents in the right side of the heart, which, as we have seen, are nearly at right angles. The fact, no doubt, bears relation to the "apex-beat" of the heart.

The arterial valves act in a simply mechanical manner. It is otherwise in the case of the auriculo-ventricular valve-apparatus, for in them the valves are dependent for their perfect function on vital contraction of muscle. The orifice which they have to close must be diminished in size by muscle contraction before they can come efficiently into play ; moreover, they must be maintained in action by the same means. "Muscle-failure" of the heart (whatever its cause) may, then, render perfectly sound auriculo-ventricular valves incompetent. In clinical medicine the importance of this fact is very great.

"Auscultatory Landmarks" now require careful consideration. Numerous observers have made painstaking investigations into this subject ; and it would seem to result from their labours that there are variations from the usual position in many individuals, allowance for which has to be made. The position of the apex and chambers of the heart has already been briefly indicated. It remains for us to determine on the surface, the points corresponding respectively to the position of the different valvular orifices, as far as in us lies, although, for the following reason, this is a matter of trifling importance :—"A superficial area of half an inch will include a portion of all four sets of valves, in situ ; an area of about quarter of an inch, a portion of all except the tricuspid."* We shall not err much in stating that the left anterior cusp of the pulmonary valves lies

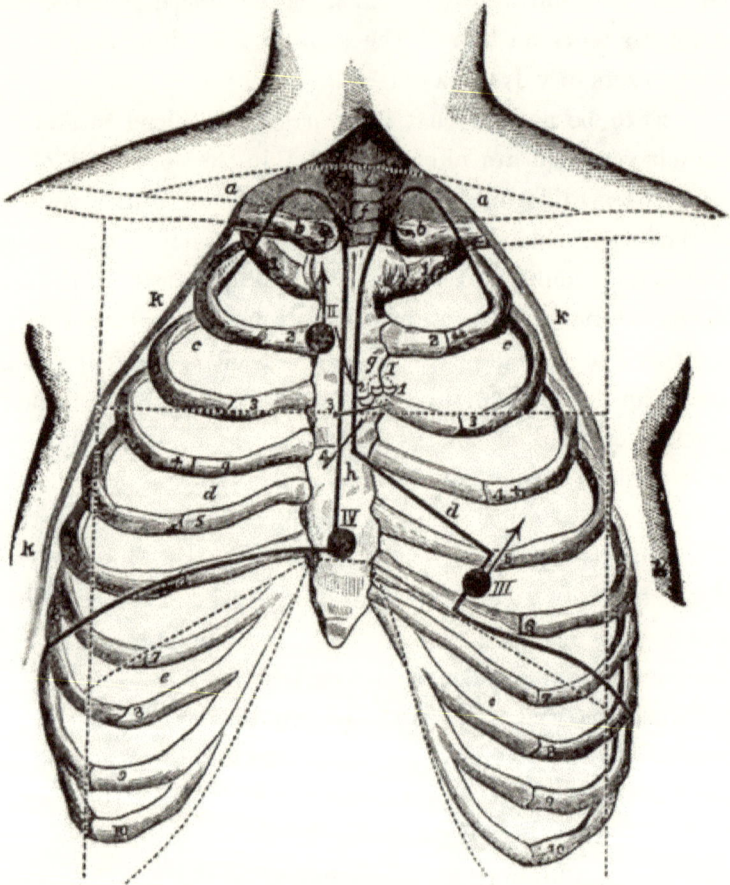

Fig. IV. represents the normal position of the different cardiac orifices, along with the areas employed for the purpose of isolating murmurs generated at these orifices respectively. 1, The pulmonary orifice (the two anterior cusps are represented) I., the pulmonary area (over the valves themselves) ; 2, the aortic orifice (the single anterior cusp is represented); II., the aortic area : 3, the mitral orifice; III, the mitral area : 4, the tricuspid orifice ; IV., the tricuspid area. The arrows show the direction of transmission of murmurs as described in the text. The arrow pointing downwards and to the left, representing the transmission of aortic regurgitation murmurs, would be better placed lower, say below the third cartilages.

behind the extremity of the third costal cartilage at its upper

* Walshe, "Diseases of the Heart," p. 6.

border, the right anterior cusp lying under cover of the sternum. Slightly lower down, to the right of, and deeper in the chest than the pulmonary valves, lies the single anterior cusp of the aortic valves. A line drawn across the sternum from the extremity of the third left cartilage in a nearly horizontal direction will represent the mitral orifice. A line drawn from the third left interspace across the sternum, and sloping downwards nearly to the fifth right cartilage, will represent the tricuspid orifice (*vide* Fig. IV.)

These considerations for ever dismiss any hope that might have been entertained as to a knowledge of the position of the valvular orifices in relation with the chest-wall enabling us by placing our stethoscope over a certain orifice to determine the production of a murmur at it. The principle upon which we are to base our method of ascertaining the orifice, at which a murmur is generated, must be other than this.

A glance at Fig. I. will show the crossing of the two great arteries, the pulmonary passing to the left, the aorta to the right; the former reaching the lower border of the second left costal cartilage, the latter the upper border of the second right costal cartilage close to the sternum. As regards proximity to the surface, the pulmonary artery is nearer the surface, opposite the third cartilage than at the level of the second cartilage, owing to this vessel passing backwards in its upward course, while the aorta approaches the surface most closely at the junction of the second right costal cartilage with the sternum. When we place our stethoscope over the sternal end of the second right costal catilage, we place it over what is termed the "aortic area," where we expect to isolate murmurs generated at the aortic orifice. The second right costal cartilage has been named the "aortic cartilage." When we place our stethoscope, again, over the junction of the third left costal cartilage with

the sternum, we cover the "pulmonary area," where we listen
for sounds generated at the pulmonary orifice.   Dr. Walshe has
named the second left costal cartilage "the pulmonary cartilage,"
but the same author admits the point here contended for.*   The
left ventricle approaches the surface at the apex of the heart,
and we listen for mitral murmurs there.   We have still to fix
an area for the tricuspid orifice, over which murmurs generated
at that orifice, have their maximum intensity.   This has been
given as "immediately above or at the ensiform cartilage,"†
again as "over the right ventricle, where it is uncovered by the
lung, *i.e.*, at the lower part of the sternum, and over the whole
space between this and the apex." ‡   Both these areas are open
to an objection, which will be afterwards explained in the section
on aortic murmurs, in which place the peculiarities of conduction
possessed by certain aortic regurgitation murmurs will be
pointed out.

The normal pericardium, closely surrounding the heart, calls
for no mention here, while the results of effusion will be best
considered when we come to "percussion," and then also the
relations of the lungs will receive attention.

footnotes

* *Vide* Walshe, "Diseases of the Heart," p. 6. par 17, and p. 98, par. 112.
† Walshe, "Diseases of the Heart," p. 95, par. 110.
‡ Gairdner, "Clinical Medicine," p. 586.

INTRODUCTORY REMARKS.

In the investigation of cardiac diseases we employ four
"physical methods of diagnosis," namely, *Inspection, Palpation,
Percussion*, and *Auscultation*.   *Mensuration* may be admitted as

a fifth. Each method rightly applied yields its own special information, and no clinical examination of the circulatory apparatus can be considered complete, which omits to make use of the data afforded by any one of them. If we regard our diagnosis as a standpoint which we take up and have to defend, we may also regard the facts ascertained by each of these methods as lines of defence with which to accomplish our object. Every line, however, is defective at some part, and if standing alone, would offer a passage to the attack of an assailant. The greatest amount of security, therefore, for our diagnosis will depend upon our having availed ourselves of the help of *all* the defensive power at our command, *i.e.*, of the facts afforded us by *all* four physical methods of diagnosis. Of course symptomatology must not be neglected, but here we have to do only with physical signs.

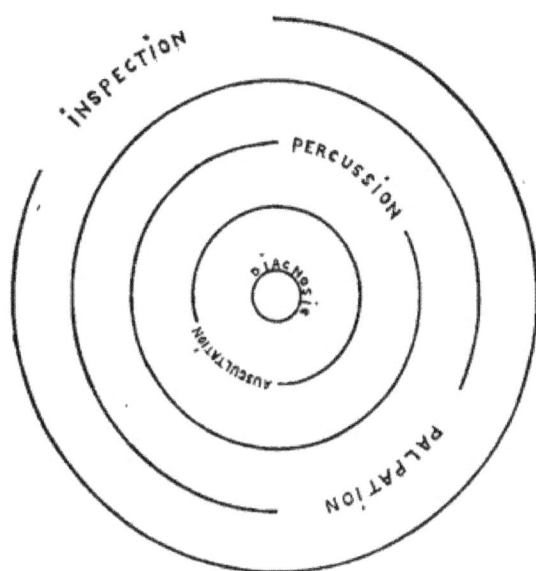

Fig. V.

INSPECTION.

It is taken for granted that we have the patient stripped before us, and usually in the recumbent position. The various points to be attended to in *Inspection* may be arranged according to the following table :—

A.

1. Bulging of the whole cardiac region.
2. Bulging of the intercostal spaces in the cardiac region.

B.—-Impulses of the Ventricles.

1. The apex-beat or impulse of the left ventricle.
   * *a.* Extrinsic displacements of the apex-beat (due to pericardial or peritoneal effusions, shrinking of the upper portion of the left lung, &c.)
   * *b.* Intrinsic displacements of the apex-beat from enlargement of the left ventricle.
2. Epigastric impulse or impulse of the right ventricle.

C.—Cardiac visible impulse other than that of the ventricles.

1. Pulsation over dilated left auricle (?).
2. Pulsation over dilated right auricle (?).
3. Pulsation over cardiac surface, which does not correspond to the impulses of the ventricles nor to the situation of the auricles.

D.—Displaced cardiac impulse (*i.e.*, when the whole organ is displaced as from pleuritic effusion, shrinking of the left lung, &c.)

E.—Vascular Pulsation.

1. Aneurism of aorta, &c.
2. Venous pulsation at root of neck.
3. Abnormal pulsation of carotids, &c.
4. Capillary pulsation.

F.

Systolic depression (1) of intercostal spaces over cardiac surface, and (2) of ribs, sternum and space between xiphoid cartilage and ribs.

A.

1. *Bulging of whole Cardiac Region.*— This occurs in great and generally chronic pericardial effusion. It is common in the cardiac enlargement following valvular disease, often accompanied by pericardial adhesion, in early life. Sometimes it exists, apparently, as a congenital peculiarity, or as a consequence of rickets. Œdema of the integuments, limited to the cardiac region, may occur in pericardial effusion of some standing.

2. *Bulging of Intercostal Spaces in Cardiac Region.*—This can only result from pericardial effusion of inflammatory origin.

B.

1. The situation of the normal apex-beat has been already indicated. Its area is small—about one square inch. When the left ventricle is enlarged, but still retains its natural form, the apex-beat becomes more extensive, and more or less altered in position.

a. *Alterations of Apex-Beat from Extrinsic Causes.*—Among the *extrinsic* causes of displacement of the apex-beat are accumulations of fluid in the pericardium and peritoneum. It is evident that ascites will also tend to tilt up the apex of the heart, causing the organ to lie more horizontally, but it will at the same time elevate the whole organ. Pericardial effusion, as a cause of altered situation of the apex-beat, is of greater interest. When fluid accumulates in the pericardium the walls of the sac yield. This they do more readily when inflamed, and we have reason to believe that it is the upper portion of the sac which gives way first, so that the fluid distends the sac upwards (*vide* Fig VII. p. 31). Whatever be the explanation of its occurrence, elevation of the apex-beat is a reliable and valuable sign in pericardial effusion. For the full value to be attached to the sign, the position of the apex-beat previous to the effusion must have been known. Shrinking of the upper portion of the left lung, as in chronic phthisis, frequently raises the apex-beat, causing the heart to lie more horizontally, or elevating the whole organ.

b. Displacement of the Apex-Beat from Intrinsic Causes.— Enlargement of the left ventricle displaces the apex-beat downwards and to the left (*vide* Fig. VI.). Enlargement of the right ventricle eventually tends to throw it to the left, but in this case the true apex-beat is apt to disappear, a more or less diffused undulatory movement between the sternum and what represents the apex, accompanied by epigastric pulsation, indicating the cardiac action. In health, if the breath be held, the apex-beat will disappear, epigastric pulsation becoming developed. This results from engorgement of the cavity of the right ventricle. In dilatation of the left ventricle in excess of hypertrophy, the true apex-beat is often replaced by a diffused pulsation, and the downward displacement of the apex is less than in cases in which the ventricle, although enlarged, retains its shape.

2. *Epigastric or Impulse of the Right Ventricle.*—Epigastric pulsation, which, we have seen, may occur physiologically, is a common phenomenon in all diseases in which the pulmonary circulation is embarrassed. In such cases the liver is frequently

enlarged from passive congestion, tending to render such pulsation more visible ; while in emphysema the depressed diaphragm is a further factor in its production. Functional palpitation of the abdominal aorta, and the pulsation of an abdominial aneurism, can hardly be confounded with the epigastric pulsation, which we are now considering, and need only be mentioned. The rare phenomenon of pulsation of the whole liver, produced by great tricuspid regurgitation, must be borne in mind.

In thus dwelling upon the preceding areas of cardiac impulse, it must not be supposed that we mean that they constitute the only seats of visible cardiac movement. They certainly are however, the important areas of cardiac impulse.

C.—Cardiac Visible Pulsation other than the Ventricular Impulses.

1. *Pulsation of Dilated Left Auricular Appendix* (?)—Occasionally, it is said, in cases of mitral stenosis the appendix of the left auricle becomes so much enlarged as to cause visible pulsation above the third cartilage to the left of the sternum, which is further stated to be systolic in time, and to be due to a regurgitant current from the left ventricle through incompetent mitral valves. In our experience systolic pulsation above the third rib, close to the sternum, is due to a dilated infundibulum of the right ventricle.*

2. *Pulsation of Dilated Right Auricle*(?)—Pulsation to the right of the sternum, auricular in origin, is a rare phenomenon, and occurs only in cases of extreme enlargement of the right side of the heart. The right auricle, which in the majority of cases alone passes to the right of the sternum, usually suffers dilatation in excess of hypertrophy ; and, where it is shown by

*In a case under my care the enlargement of this portion of the right ventricle was so great as to give rise to extension of the cardiac dulness in the upward direction, simulating the dulness of pericardial effusion.

percussion to be much enlarged and even to have pushed aside the superimposed lung, no pulsation may be visible. When pulsation is visible over the right auricle it is probably produced by the mechanism described above with reference to the left auricle, and is therefore systolic and not presystolic.

When there is sufficient enlargement of the right auricle to be detected by percussion there is probably always a condition of *systole catalectic*, or imperfect systole of the auricle.

3. *Pulsation over the Cardiac Surface, which Pulsation does not correspond to the impulse of the ventricles nor to the situation of the auricles.*—As already indicated, when the right chambers of the heart are enlarged, pulsation may become visible to the left of the sternum in the second intercostal space, the pulsation corresponding to the situation of an enlarged infundibulum of the right ventricle. In cases of intra-thoracic tumour in the posterior mediastinum, in which the heart is pushed forwards from behind, the cardiac impulse is observed in an area almost coextensive with that of the whole organ, and the usual impulses of the right and left ventricles escape recognition in the diffuse general impulse. The heart during its diastole accommodates its shape to surrounding parts, but during systole it asserts the circular outline presented by its ventricles in a state of contraction, hence the forward impulse, when the posterior mediastinum is encroached upon by tumour.

D.

By displaced cardiac pulsation is meant pulsation of the heart, *away from* the normal cardiac region, *with absence of pulsation in the normal area*, not merely pulsation due to extension of the normally situated heart beyond its legitimate limits. Displacement of the heart is most commonly observed in left pleuritic effusion, in which case it may lie quite to the right of the sternum. Retraction of lung substance may draw the heart to the right or left, but generally upwards, inasmuch as it is the upper portions of the lungs which chiefly undergo phthisical contraction. Ascites, ovarian cystic disease, and tympanitic distension of the bowels, push up the whole heart, besides causing it to lie more horizontally. Hydatids growing from the upper surface of the liver likewise displace the heart. Aneurisms of the arch of the aorta may again depress the heart, while mediastinal growths may push it aside according to the direction of their development.

E.

1. A pulsating area above the third rib, the pulsation being systolic in rhythm, and the apex-beat of the heart not having undergone elevation, is probably vascular in origin, but if close to the sternum it may be due to a dilated infundibulum of the right ventricle, or possibly to a dilated left auricle. If the pulsation be over a wide area and heaving, (*vide* "Palpation"), it is aneurismal. In certain deformed chests, even though the deformity be not extreme, an abnormal position of the arch of the aorta may closely simulate an aneurism. In contraction or collapse of the left lung, a deceptive appearance of aneurism may be given, from exposure of the pulmonary artery, frequently in such cases dilated. Aneurisms of the arch of the aorta generally become visible above the third rib, to the right or left of the sternum, or in the middle line, this bone having been corroded by pressure. Great care should be taken to have the patient in a good light and to observe the chest from different points of view. It is especially important to look along the surface of the suspected area.

2. *Venous Pulsation at Root of Neck.*—This is an important feature of cardiac disease. It is almost necessarily accompanied

by abnormal fulness of the veins. The right side is usually
the more, it may be only, affected. We must remember that
respiration exerts a powerful influence on the return of blood to
the thorax. Thus, during an expiratory effort, we see the veins
of the neck swell up; while, during a deep inspiration, the
circulation in them is accelerated, and they collapse. In cases
of adherent pericardium, sudden collapse of the veins during the
cardiac diastole has been observed. This was supposed by
Friedreich to be due to stretching of the intra-thoracic veins by
the heart adherent to the diaphragm and chest-wall, during its
systole, and subsequent recoil during the diastole. Venous
pulsation is auricular-systolic or ventricular-systolic, or both,
but the former is of less magnitude and importance.
The tracing of a jugular pulse affords evidence of an
auricular contraction-wave preceding the ventricular-wave.
Distension of the veins, if habitual, as in cases of emphysema
and cardiac disease, must render the valves of these vessels incompetent; and it is probable that the phenomenon which we
are considering is never developed in perfection without the
occurrence of this defect. Much care is necessary on the part
of the beginner to distinguish true venous from communicated
arterial pulsation. The finger should be pressed lightly upon
the external jugular at the root of the neck; if the pulsation
continue, it is evidently arterial. Moreover, in the case of
the venous pulse, by pressing a finger of the other hand
upwards and emptying the vein from below, on removing the
finger from the root of the neck, the blood will be seen to
rush up according to the degree of incompetence of the valves
of the veins and tricuspid orifice. When well marked, jugular
pulsation may be regarded as a clear indication of incompetence
of the tricuspid valves; but it must be remembered that
tricuspid incompetence may exist without this sign, if the

venous valves remain competent. This is most likely to occur where there is only temporary engorgement of the right cardiac cavities. On the other hand, a slight degree of pulsation may arise simply from the percussion of the tricuspid closure, reacting on the blood column in over-full veins.

3. Abnormal visibleness of arterial pulsation accompanies one cardiac valvular lesion as a peculiar feature. This valvular lesion is aortic regurgitation. Mere exaggerated visibleness hardly adequately expresses the peculiar jerk which characterises the pulse in the affection named. As this subject, however, will again, require our careful attention, we may be brief here. There is one peculiarity of aortic regurgitation which is best appreciated by inspection—namely, a locomotion of the vessels forwards, and, if they be tortuous, in somewhat vermicular fashion. In all cases in which the left ventricle is enlarged, carotid pulsation tends to become unduly visible; so also in cases without cardiac enlargement, whenever the circulation is excited temporarily by emotions, &c., or permanently, as in Graves' disease. Morbid changes (atheromatous) in the arteries themselves likewise simulate in some respects the pulse of aortic reflux; but with the precaution to fix our attention on the larger trunks, the sign under consideration is of great diagnostic value.

4. Another phenomenon, which, when well marked, is almost characteristic of aortic reflux, is the so-called "capillary pulsation." By rubbing the finger nail over the skin, say of the forehead, a red mark is produced, and after a few seconds, if this be carefully watched (it is often most evident some little distance off), a distinct wave of deepening and paling in colour is observable over the surface of the artificial erythema, while its edge extends and recedes.

c

F.

1. *Systolic Depression of the Intercostal Spaces in the Præ-cordial Region.*—This is not uncommon in the fourth space (and even third) above the apex-beat, when the heart is enlarged, or at all events comes into more extensive contact with the chest-wall. There is in such cases still an apex-beat.

2. *Systolic Depression of the Ribs, Sternum, and Space between Xiphoid Cartilage and Ribs, accompanied by absence of the Apex-Beat.*—This indicates universal pericardial adhesion, the præ-cordial pleural layers being likewise united, but it is a rare phenomenon. Most cases of simple pericardial adhesion cannot be recognised by physical signs, although the condition is one of great importance in cardiac pathology, inasmuch as it exerts an injurious influence on the heart-muscle.

PALPATION.

Signs which may be ascertained by palpation :—

A.—Cardiac Impulse.

1. Situation and extent of apex-beat.
2. Strength of apex-beat.
3. Strength of epigastric impulse.
4. Closure of semilunar valves with undue force perceptible to hand.
5. Diastolic backstroke.

B.—Vibrations perceived by means of the hand applied over Cardiac Region.

1. Thrills { Presystolic. Systolic. Diastolic.
2. Friction fremitus.

C.—Thrills, &c., perceived by means of the hand placed over Pulsating Areas other than that of the Heart.

1. Aneurisms.
2. Dilatation of aorta, or normal aorta unduly in contact with chest-wall.

In the use of palpation as a method of cardiac examination, it is important that the whole hand should be placed flat upon the surface, and this is especially necessary when the apex-beat has to be sought for.

A.

If the apex-beat be invisible, its position must be ascertained, if possible, by palpation. The normal situation of the apex-beat has already been defined, and some of the changes it undergoes in disease indicated. When inspection and palpation alike fail to reveal the presence of an apex-beat in the recumbent posture, the patient should be asked to sit up, inclining forwards (*not inclining to the left side*, if we would determine the *position* of the apex). The absence of the apex-beat becomes an important item in the data for a diagnosis, due allowance being made for unusually thick chest-walls, emphysema, the apex beating under a rib, &c., and care expended on the investigation is not lost, though the object of our search be not found.

While any discussion as to the mode of production of the physiological apex-beat would be out of place here, the fact of the apex-beat being the expression of the contraction of the left ventricle must be emphasized. When this chamber undergoes enlargement, the apex-beat is necessarily altered in situation, and it becomes more extensive. The presence of a true well-defined apex-heat implies the retention, in greater or less degree, by the ventricle, of its normal pointed form. When the chamber becomes dilated and rounded, the true apex-heat is lost, and only a diffuse impulse remains to represent the systole of the ventricle, or there may be no impulse perceptible, although the ventricle is enlarged. The vigour of the heart-muscle and the degree of effort, with which the left ventricle accomplishes its contraction, are elements which exert a great influence on the production of the apex-beat. In severe fever, for instance, the apex-beat may be watched failing day by day until it is entirely lost. In Bright's disease, with cardiac hypertrophy well maintained, the apex-beat is slow and heaving, while in palpitation of nervous origin the apex-beat is quick and slapping. Certain valve lesions are accompanied commonly by peculiarities of the apex-beat. Thus aortic incompetence is generally accompanied by a vigorous apex-beat so long as the heart-muscle remains healthy. There is some reason to believe that the apex-beat may fail in great aortic obstruction, although the left ventricle is hypertrophied and its muscle sound. Many cases of mitral stenosis are accompanied by a well-marked apex-beat, which corresponds with the sharp abrupt first sound, that is characteristic of the lesion. In this case it may be supposed that the ventricle begins to contract before it has been distended to the full amount—before, in fact, the current from the auricle has ceased.

The impulse of the right ventricle is altogether different from the apex-beat, and seems to be delivered by the broadside of the ventricle. In this relation the position of the pulmonary artery with regard to the right ventricle may be borne in mind. Epigastric impulse is, at best, only an indirect impulse, conveyed through the liver, and it is always diffuse and ill-defined. The apex-beat and the epigastric impulse generally vary inversely—when the one is pronounced the other is defective or absent. Thus, when there is much enlargement of the right side of the heart, the left ventricle is pushed back by the enlarged chamber, and in long-standing cases the right ventricle reaches the apex of the heart. In certain cases of aortic regurgitation, again, the left ventricle for long is able to cope with the difficulty, so that the embarrassment of the circulation does not pass backwards through the lungs to the right side of the heart. Such cases have a very exaggerated and displaced apex-beat, but little or no epigastric impulse, while engorged liver and dropsy are conspicuous by their absence.

It is seldom that both impulses—those of the left and right ventricles—are well marked at the same time.* In severe fever (as Typhus) while the apex-beat fails, epigastric impulse is established and increases, as the result of the obstruction in the pulmonary circulation, invariably more or less present under the circumstances, and of the retention of vigour in greater degree and for a longer period by the right ventricle in comparison with the left ventricle. When the left ventricle becomes dilated and rounded, its systole is invariably imperfect—*systole catalectic*—and whether there be or be not mitral regurgitation, the right side of the heart becomes involved in the obstruction suffered by the circulation. Thus we have two causes at work in obliterating the apex-beat and establishing or exaggerating epigastric impulse. While the apex-beat is a normal phenomenon, epigastric impulse is generally, though by no means always, abnormal.

1. *The Situation of the Apex-beat.*—When the left ventricle is enlarged and retains its pyramidal form the apex-beat is displaced downwards and towards the left. Its extent, too, normally about a square inch, is increased in greater or less degree.

In pericardial effusion the apex-beat is elevated. Other circumstances influencing the position of the apex-beat, have been already referred to under "Inspection."

2. The strength of the impulse is a matter of much importance both in cardiac and general diseases. Of the former, fatty metamorphosis and the compensatory hypertrophy of aortic-valvular disease offer good examples. Of the latter, the weakening of the heart in typhus, so admirably described by

---

* Some time ago a patient was admitted into the Manchester Infirmary under my care, with typical emphysema of the lungs. But in addition to epigastric impulse, he presented a very well-defined vigorous apex-beat: on careful consultation, a faint blowing diastolic murmur of aortic incompetence was heard, which fully explained the condition of the apex-beat.

Stokes, and the hypertrophy consequent on Bright's disease, furnish illustrations. In all these cases, it is the left ventricle which is first and chiefly affected, and consequently it is the apex-beat, which undergoes alteration. Pericardial effusion diminishes the force of the apex-beat.

3. *Epigastric Pulsation.*—This phenomenon is better perceived by "Inspection," but in cases in which there is great enlargement and hypertrophy of the right ventricle, a distinct heaving is felt over the lower part of the sternum and in the epigastrium..

4. *The Closure of the Pulmonary Semilunar Valves* is frequently perceptible as a shock to the hand placed on the surface of the chest over the appropriate area, in cases in which there is obstruction to the pulmonary circulation from any cause. It will be rendered more distinct in proportion as the pulmonary artery is exposed by retraction of the lung. The pulmonary orifice, it will be remembered, is situated near the surface of the chest. A similar phenomenon may be perceived over the aorta in cases of dilatation of the vessel with high arterial tension.

5. *Diastolic Back-Stroke.*—This is only rarely met with as a sign of adherent pericardium with cardiac enlargement. My experience of the phenomenon would lead me to regard it as resulting from the rebound of the dilating heart after the systolic depression already noted under "Inspection" (F. 2).

B.—Vibrations perceived by means of the hand applied over the Cardiac Region.

These naturally fall into two classes—endocardial and exocardial.

1. *Endocardial Thrills.*—Thrills are produced at or rather beyond the cardiac orifices by the *veines fluides* which give rise

to the corresponding murmurs. A thrill is most common where there is obstruction to the blood-current at a rigid and contracted orifice, as in aortic and mitral stenosis.

The aortic obstructive thrill is systolic, the mitral obstructive thrill is presystolic or diastolic in rhythm. In case of dilatation of the aorta, a systolic thrill may be perceptible over the vessel, although there is no actual constriction of the orifice. The presystolic thrill of mitral constriction, which accompanies the systole of the auricle, is limited to the apex, and possesses the peculiar feature of increasing in intensity up to the apex-beat, which at once cuts it short. The diastolic thrill of the same lesion accompanies the expansion of the left ventricle, and diminishes in intensity towards its end. It also is an apex thrill. When they co-exist, as they often do, these two thrills can be easily distinguished. Sometimes mitral regurgitation is accompanied by systolic thrill at the apex, but this is comparatively rare. Aortic regurgitation is more frequently attended by diastolic thrill, felt at the apex of the heart, to which it may be limited. The incompetence of the valves is usually considerable in cases of the kind. In certain rare cases of very free aortic regurgitation, to be referred to later, a short indistinct thrill, apparently of presystolic rhythm, may be perceptible at the apex, and simulate, in some degree, the auricular-systolic thrill of mitral stenosis.

2. *Friction Fremitus*, due to exudation on the pericardial surfaces, need be here only mentioned, as the subject of friction is fully considered in the section on Auscultation.

C.—Thrills, &c., perceived by means of the hand applied over Pulsating Areas other than that of the Heart, viz., Aneurisms.

1. When an impulse, other than that of the heart, yet

synchronous with its systole, is at least as forcible as the cardiac pulsation, very reliable evidence is afforded of the presence of an aneurism. The murmurs audible over aneurisms are sometimes represented in palpation by thrills. Again, as already indicated, when the semilunar valves are closed with abnormal force, as is always the case in aortic aneurism, a distinct thud may be felt by the hand placed over the tumour.

2. When the aorta is dilated and in exaggerated contact with the chest-wall from any cause, a thrill may be perceptible over the area of contact. Such thrills accompany murmurs, and their causes will be considered under "Auscultation."

PERCUSSION.

I.—"Superficial" Dulness indicates the extent of cardiac surface in direct contact with chest-wall.

II.—" Deep " Dulness indicates approximately the absolute size of the heart.

In front the heart being surrounded, except inferiorly, by air-containing lung, we are enabled to estimate its size or the amount of pericardial effusion, by means of percussion, inasmuch as the heart or distended pericardium, yields no resonance over that portion of the chest-wall with which it is in contact, and modifies the resonance of the lung beyond that area by rendering the layer of resonant tissue superficial to the heart less thick.

Percussion determines for us the size of the heart or pericardial sac distended with fluid, also the area in which aneurismal tumours come into contact with the chest-wall. In health the large vessels cannot be recognised by percussion. If they could they would form a more or less circular dull area, super-imposed upon the cardiac dulness proper, *i.e.*, above the level of the third rib (*vide* Fig. I.).

There are two methods of ascertaining the size of the heart by percussion. One is to determine what is called "the superficial cardiac dulness." This really means mapping out the area in which the heart comes into contact with the chest-wall. The anterior border of the left lung separates from the anterior border of its fellow at the fourth rib, and passes outwards and downwards to about the union of the fifth rib with its cartilage; it then extends downwards and inwards along the sixth left cartilage, thus leaving a rudely triangular space in which the heart (chiefly the right ventricle) comes into contact with the thoracic wall (*vide* Fig. I.). If this space formed a regular triangle, the superior or external side would measure three inches, the internal side corresponding to the middle line, two inches, and the inferior two and a half inches, but it is only approximately of triangular form. Except at the right side, where *sternal* resonance modifies the percussion note, this dulness is easily defined by light percussion. The peculiar resonance of the sternum interferes to some extent with our determination of the right border of cardiac dulness, unless the right auricle is engorged, and the dulness extends to the right of the bone. But it is often possible to detect a decided difference in the percussion sound of the right and left halves of the sternum at the level of the heart. The information acquired by this mode of percussing the heart, though alone relied upon by many physicians, is open to the objection that it does not tell us the absolute size of the heart, but only the size of the cardiac surface which is in contact with the chest-wall.

The other method of estimating the size of the heart by percussion depends upon the modification which the organ produces on the resonance of the lung overlying it: it determines approximately the actual size of the heart. In performing percussion according to this method, we have, as it were, to percuss through a certain depth of lung-tissue. A few experiments in the *post-mortem* room will make abundantly plain the fact, that a thin layer of lung tissue gives a "less full" or "more empty" note than a thick layer. In percussing from well beyond the cardiac region towards the triangular area of superficial cardiac dulness, the gradual transition from the full pulmonary note elicited beyond the heart, to the complete emptiness or dulness of the percussion note in the triangle referred to, in proportion as the heart encroaches on the lung space, will be easily appreciated. Until the heart comes into immediate contact with the chest-wall, the lung will yield its clear and resonant note, though one becoming more and more "empty." By practice the power of determining sufficiently for all clinical purposes the line at which the heart begins to modify the pulmonary note is soon acquired.

In cases of "large-lunged emphysema," thick masses of emphysematous lung-tissue spread over the heart, so as to render this method of percussion useless. But, in this case, the other or superficial method is of no greater value, as the voluminous left lung encroaches upon the triangle.

With regard to the deep method, percussion in two directions is sufficient for ordinary purposes. These are, in a line drawn vertically downwards one inch to the left of the left border of the sternum, and in a line drawn transversely along the fourth rib, or in the male, through the nipple (*vide* Fig. VI.). The normal limits are as follows :—Dulness commences at the third rib and terminates at the sixth rib, but this latter limit cannot be ascertained satisfactorily, owing to the junction of cardiac and hepatic dulness. Percussing from left to right, dulness should commence just within the nipple line, and should not transgress the right border of the sternum. Though in health the right auricle reaches to half an inch to the right of the right border of the sternum, it lies beneath the comparatively thick border of the right lung, and cannot be distinguished by percussion. It is to be noted, however, that there is often some slight diminution of resonance for barely a finger's breadth along the right sternal margin at the level of the right auricle, in cases in

which there does not seem to be, from other evidence, any
enlargement of the auricle. This may be due to approxima-

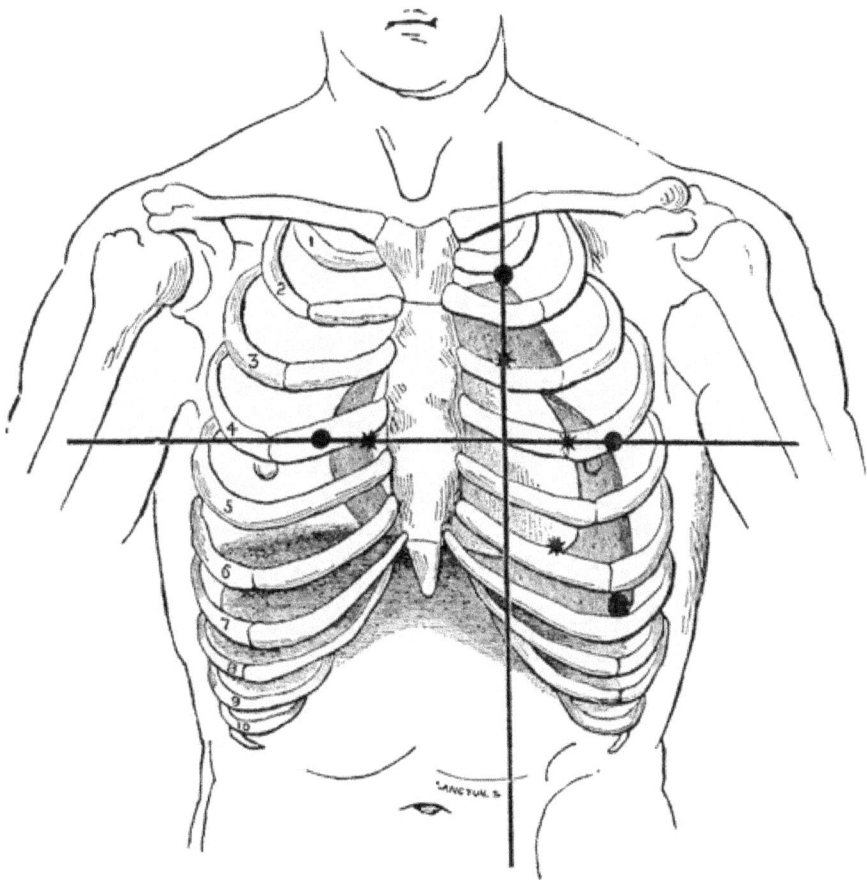

Fig. VI. represents the method of percussion described in the text. The clear space
indicates the normal anatomical area of the heart, the asterisks marking the spots
where percussion, in the directions indicated, first detects the influence of the under-
lying solid organ upon the lung resonance (as mentioned in the text, in health
dulness is hardly ever detectable to the right of the sternum). The dots represent
the boundaries of abnormal dulness :—Extension upwards, indicating pericardial
effusion. Extension to the right, indicating enlargement of the right auricle.
Extension to the left, indicating enlargement of the left ventricle. The normal
position of the apex-beat is represented by an asterisk. The depressed apex-beat by
a dot. The inferior extremity of vertical dulness is unsatisfactory. Valuable
information as to the lower limit of the heart may be obtained from taking into
consideration the upper border of the liver dulness to the right of the sternum.

tion of the cartilages of the ribs as they join the sternum. By
inspection and palpation we are able in the great majority of
cases to determine the position of the apex-beat, and by means
of it we are enabled, with the limits of vertical and transverse
percussion just described at our disposal, to make a rough
outline sketch of the cardiac organ in the following way (*vide*
Fig. VI.) :—A curved or semi-lunar line is to be drawn from the
apex-beat to the upper extremity of the vertical dulness in the
parasternal line (*i.e.*, one inch to the left of the left border of the
sternum), which line should pass through the left extremity of
the transverse dulness. Again, a similar curved line is to
be drawn from the apex-beat to the right extremity of the
transverse dulness. Lastly, the two lines drawn from the
apex are to be joined by a third curved line. In this way a
rough outline of the heart is obtained. Measurements can be
made, taking the middle line of the sternum as the fixed point
for those in a transverse direction to the right and left, and the
cartilages of the ribs as the fixed points for longitudinal measurements. The upper border of hepatic dulness to the right
of the sternum should always be determined. The indications
afforded by abnormities in the cardiac dulness are briefly as
follows, percussion being made in the manner described :—

(1.) Increase upwards = pericardial effusion (rarely enlargement
of the infundibulum of the right ventricle, possibly enlargement of the appendix of the left auricle).

(2.) Increase to the right = enlargement of the right auricle.

(3.) Increase to the left = dilatation of the left ventricle. (But
great enlargement of the right ventricle ultimately leads to the
same result).*

---

* In cases in which emphysema seems to be the primary mischief, it is no uncommon
thing to find both right and left ventricles of the heart enlarged. It seems more likely,
in these cases, that there is a degenerative tendency which has manifested itself in the
heart as well as in the lungs than that the dilatation of the left ventricle is sequential
to that of the right.

As already mentioned, increase upwards is the chief characteristic of the dulness of pericardial effusion, but a hardly less
marked feature is the form assumed by the dull area (*vide* Fig.
VII.). Only when the case is chronic does the distended pericardium assume a more globular form.

When the apex of the heart is found beating considerably
within the dull area, almost conclusive evidence of pericardial

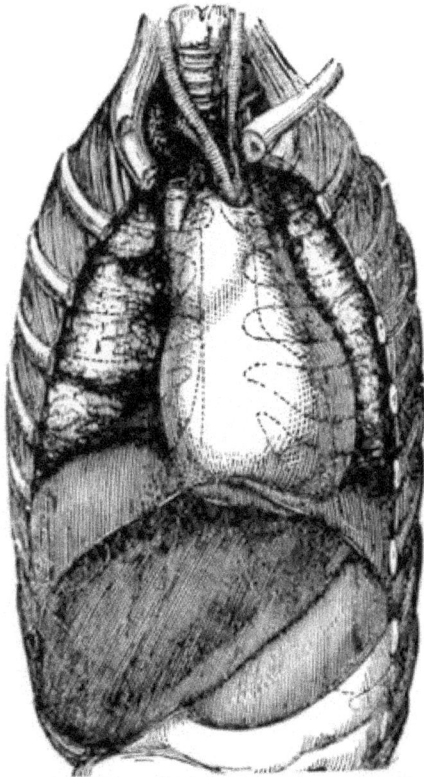

(Sibson).

Fig. VII.—In this figure the effect of effusion into the pericardial sac is shown. The
cardiac (?) dulness will increase upwards, and the dull area assume more or less the
shape indicated here anatomically, being broadest inferiorly.

effusion is afforded, the lungs being, of course, resonant in the
neighbouring regions. In simple dropsy of the pericardium, the
pyramidal shape assumed by the dulness is less marked.

The determination of the size of the heart or of a distended pericardium, by percussion, may be interfered with by pleuritic effusion, or consolidation of the neighbouring portions of lung, but these usually present no difficulty in diagnosis, as characteristic pulmonary physical signs are pronounced. Truly disheartening, however, to the beginner, are the results of what Dr. Walshe has well called "horizontal conduction." For instance, in a case of cardiac disease with enlargement of the organ, the heart is seen pulsating over a wide area, yet over the

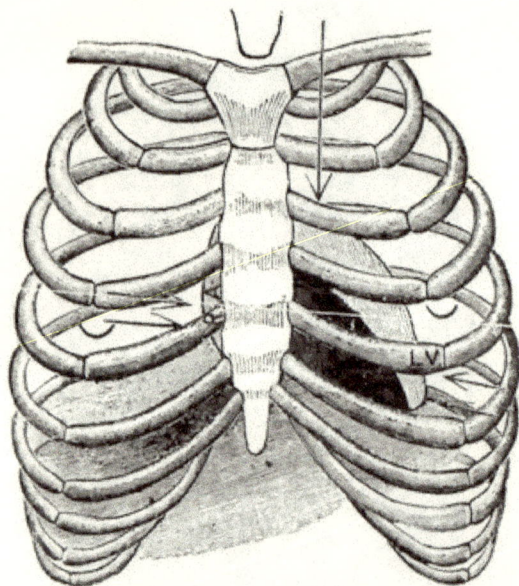

Fig. VIII.

very pulsating area a tympanitic note is elicited on percussion —a distended stomach lying beyond, being the cause of the anomaly. This is an example of the importance of the rule never to trust to *one* physical sign. Even in these circumstances the *lightest possible* percussion will generally yield dulness over the part of the chest-wall in contact with the heart.

For purposes of practical diagnosis, the essential points to be ascertained are the following :—(1) The size of the right auricle, (2) the size of the left ventricle, and (3) the presence or absence of pericardial effusion. This may be done by a few very simple manipulations.

Percuss downwards an inch or more, if necessary, to the right of the sternum until the liver dulness (the lung-margin) is reached. Mark the spot with a copying-ink pencil. Then percuss towards the sternum not less than an inch above the level marked, and note the spot at which the modification (if any) of the lung sound, due to the auricle beneath, becomes appreciable. When absolute dulness to the right of the sternum is detected in this way, it signifies a very great degree of distension of the auricle, which has displaced the lung margin. Such distension, no doubt, implies imperfect systole—*systole catalectic*—on the part of the auricle. It is a good habit to compare the percussion sound to the right of the sternum above and below the third cartilage, when any modification of the lung-sound has been noticed at the latter level.

We next turn our attention to the left of the sternum and endeavour to determine the size of the left ventricle. Inspection and palpation having preceded percussion, the situation of the apex-beat will probably be known. If so, let us mark the spot, noting the intercostal space in which it is situated. This spot will correspond pretty closely to the level of liver dulness already marked on the right side, or it may be a little lower. Let us next percuss from the anterior border of the axilla, or, if necessary, further outwards, towards the heart, choosing the level of the apex-beat if that has been determined by inspection or palpation, or if not, the level corresponding to rather less than an inch above the liver dulness, to the right of the sternum. Whether the situation of the apex-beat has or has not been

D

revealed by inspection or palpation, the cardiac dulness reached at the level indicated will represent the apex of the heart closely enough for all practical purposes, and the apex is the part of the heart which extends furthest to the left. Outwards (*i.e.*, to the left) and downwards are the directions in which the apex of the heart is displaced, when the left ventricle is enlarged, and it fortunately happens that when the downward displacement is most pronounced, the apex-beat is usually easily discerned—the ventricle retaining more or less its normal form instead of becoming rounded—so that any error that might arise from our getting above the apex is prevented. It is to be remembered in relation to this method of determining the size of the left ventricle, that approaching the apex of the heart we very soon reach "superficial" dulness (*vide* p. 1).

The upper boundary of the heart-dulness is determined by percussion in a vertical direction, an inch to the left of the sternum. Normally, as already stated, there should be no modification of lung-resonance due to the heart above the third left cartilage.

The results of percussion of the heart in a given case may be shortly noted in the following way:—The left cartilage to which the cardiac dulness (in the sense of lung resonance modified by the heart) reaches, is recorded in Roman numerals III. or II, as the case may be, while the extension of dulness to the right and left respectively from the middle line is noted in Arabic figures placed below. Thus, $\frac{\text{III.}}{3-7}$ was noted in a case of aortic regurgitation, in the last stage of the disease, and implied enormous enlargement of the left ventricle and great distension of the right auricle, with absence of pericardial effusion. In this disease, enlargement of the right side of the heart may be long delayed. The figures $\frac{\text{III.}}{1\frac{1}{2}-4}$ were noted in a case of mitral stenosis and indicated some distension of the right auricle, slight enlargement of the left ventricle, and absence of pericardial effusion.

Good practical rules for guidance are, that normally (1) there should be no dulness to the right of the sternum, and that (2) the cardiac dulness to the left of the sternum should not extend beyond a line drawn vertically downwards from the nipple, and that.(3) modification of resonance due to the heart should not be encountered above the third cartilage, when percussion is made in a vertical direction an inch to the left of the sternum.

When it is remembered that the heart is a contractile organ, constantly undergoing changes in size and shape, the utility of attempting very accurate measurements of it by percussion, may well be doubted. On the other hand, however, it is certain that an approximate estimation of the size of the heart can be made by means of percussion, and of all the physical signs afforded by the heart there is none of greater—perhaps of so great —value as that which indicates enlargement of the organ as a whole or in part. The size of the right auricle enables us to gauge the obstruction in the pulmonary circulation and more or less in the general circulation. When there is enlargement of the right auricle, confirmatory evidence of such obstruction will not be wanting—dyspnœa, enlarged tender liver, dropsy, &c. Enlargement of the left ventricle depends, for the most part, upon dilatation of its cavity—an unmixed evil, unlike hypertrophy of its walls. As a matter of clinical and pathological experience it will be found that valve-lesions which tend to induce excess of pressure within the chamber *during diastole* are most potent in the production of dilatation. Aortic regurgitation offers the best example of this fact. In pure aortic obstruction it may be doubted, if dilatation of the left ventricle ever supervenes, before the ventricle has assumed from failing power, the condition, which we have called *systole catalectic.*

In this relation it may be permitted to quote from a recent work of Drs. Roy and Adami on "Heart Beat and Pulse Wave," as follows:—"One of the effects of rise of pressure in the systemic arteries is to diminish the extent to which the fibres of the heart-wall are shortened during systole . . . . The effect of the diminished shortening is, as we have shown, to increase the quantity of residual blood which is left in the ventricle at the end of systole. This increase in the residual blood does not, however, under normal conditions, lead to a diminution of the amount of blood expelled by the heart in a given time, seeing that it is compensated for by an increased expansion during diastole."

AUSCULTATION.

A. (*a.*) The normal heart sounds.

    (*b.*) Modifications of these.

      1. Intrinsic (accentuation, reduplication, weakening, &c.)

      2. Extrinsic (effects of pericardial effusion, obesity, &c.)

B. New or adventitious sounds.

I. Endocardial Murmurs.

(*a.*) Valvular, *i.e.*, formed at or beyond the auriculo-ventricular or arterial orifices from structural changes in the valves or orifices, or from the auriculo-ventricular valves being rendered incompetent by muscle-failure, or from alteration in the relation of the size of an arterial orifice to that of the adjoining vessel.

    b. Anæmic murmurs, &c.

(*a*) Obstruction $\left\{ \begin{array}{l} \text{absolute.} \\ \text{relative.} \end{array} \right.$

(*β*) Regurgitation.

    1. From disease of valves.

    2. From muscle-failure (auriculo-ventricular valves).

    3. From dilatation of orifice (arterial valves).

Murmurs their $\left\{ \begin{array}{l} \text{1. Rhythm.} \\ \text{2. Maximum intensity.} \\ \text{3. Direction of transmission.} \end{array} \right.$

II. Exocardial Murmurs or Friction $\left\{ \begin{array}{l} \text{Pericardial.} \\ \text{Pleural (of cardiac} \\ \text{rhythm).} \end{array} \right.$

Vascular Sounds.

In health.

In disease.

    1. Arterial.   3. Aneurismal.

    2. Venous.

A. (*a.*) The cause of the normal "sounds" of the heart is a subject which cannot be altogether omitted, inasmuch as the inferences to be drawn from the abnormal sounds heard in disease will to some extent depend upon the mechanism assigned to each. For practical purposes the "valvular" theory, which regards the sudden closure of the valves as the main element in the production of the "sounds," is satisfactory enough. For practical purposes it is necessary, however, that the auscultator should bear in mind various factors which exert an influence upon the heart sounds in one way or another, without his necessarily believing in a "muscle-contraction" sound.

We must recognize a "muscle contraction" factor in the production of the first "sound" as heard at the apex. With regard to it, rapidity of contraction would seem to play an important part; possibly there may be an "arterial wall" element in the "sounds" as heard at the base of the heart over the great arteries. The fact that a systolic sound closely simulating a "sound" of the heart can be heard in the femoral artery in cases of aortic regurgitation, in which the "sounds," being replaced by murmurs are inaudible over the heart itself, is matter of common observation. Moreover Traube showed that in certain cases of extreme aortic regurgitation actually two "sounds" may be heard over the femoral artery, although the mode of production of the latter has not as yet been explained.

In clinical descriptions, "sounds" *in the technical sense* must be carefully distinguished from "murmurs." In the cases referred to Traube drew attention to this distinction, as the very rare phenomenon observed by him had been confused with the common sign of a double *murmur*, in aortic regurgitation. A "sound," resembles a normal cardiac "sound," Lubb or Dup ;

murmurs, on the other hand, are more or less prolonged and blowing. Pericardial Friction should not be called a murmur, and gives to the ear an idea of the mode of its production. hardly to be mistaken.

The condition of the valves seems, in the absence of murmurs, to have some influence on the cardiac "sounds." For instance, in cases of great aortic stenosis from rigid valves having only a slit between them, neither second "sound" nor diastolic murmur may be audible. It has been supposed by some, that the common funnel-shaped deformity of the valves in mitral stenosis is a factor in the production of the characteristically accentuated first sound of the lesion. The first "sound" of the heart corresponds to the commencement of the contraction or systole of the ventricles and to the closure of the mitral and tricuspid valves, the second "sound" to the commencement of expansion or diastole of the ventricles, and of the recoil of the great arteries, and to the closure of the arterial valves. How much "sucking" force may be exercised by active expansion of the ventricles it is difficult to estimate ; but there are reasons for believing that such force may contribute materially to the production of certain murmurs (mitral diastolic). We must remember that the "sounds" are not co-extensive in time with the systole or diastole of the ventricles, inasmuch as a mitral regurgitation murmur is frequently heard along with the first "sound" *and following it*, showing that the contraction is going on after the "sound" has terminated, while a diastolic murmur is often heard following the second sound.

The right and left sides of the heart each produce two "sounds,"; and although we hear in health only two "sounds," the occurrence of non-coincidence in time of the two pairs of "sounds" might be conjectured under morbid conditions (*vide*

Reduplication of " Sounds "). These and other points will call for further notice as the various abnormities met with in disease are considered in detail.*

(b.) Modifications of the Heart Sounds.

1. *Intrinsic Modifications—Accentuation.*—This character is commonly observed in the second sound. As the name indicates, by accentuation is meant an intensification of the sound, owing, apparently, to the closure of the valves taking place with abnormal force. In all valvular diseases of the heart accentuation of the second sound over the pulmonary artery is the rule, and, indeed, must occur whenever there exists hindrance to the pulmonary circulation. Regurgitation through the tricuspid orifice will, of course, tend to diminish the sign. In cases of Bright's disease, in which there is high arterial tension, the same character is given to the aortic second sound, and most markedly when there is dilatation of the aorta. In cases of thoracic aneurism, in which the diagnosis is difficult, accentuation of the second sound over the aorta may prove a useful indication. The first sound is said to be accentuated, when it is short, abrupt, and "thumping," *i.e.*, accompanied by a distinct shock. The loud abrupt first sound of mitral stenosis may be explained by the existence of difficulty in filling the left ventricle, due to the lesion, so that the ventricle contracts before it is fully distended, and, therefore, has a light burden, so to speak, while its nutrition is unimpared. The rapidity of contraction has probably much to do with this character of the sound. In hypertrophy of the left ventricle an opposite character is assumed by the first sound ;

---

* In the preceding remarks the word *sounds*, meaning the normal cardiac sounds or sounds of the same character, has been distinguished by inverted commas. This will not be done in the rest of the work, as the significance of the term in its technical sense has been explained, and no doubt is likely to arise as to the sense in which the word is to be taken, the general or technical.

it becomes dull and toneless. For an explanation we must have regard to the cause of hypertrophy : it is, shortly, increased work calling forth *effort*. Every contraction of the ventricle under these circumstances implies abnormal resistance overcome. In certain cases of dilatation of the heart, on the other hand, the first sound is short and peculiarly clear. A fair degree of nutrition of the ventricular walls seems necessary for the production of an accentuated first sound. In fevers the modifications undergone by the heart sounds are of peculiar interest, and were admirably described by Dr. Stokes. The first change is shortening of the first sound, so that it comes to resemble the second sound. This stage is accompanied by low arterial tension, while the ventricular walls have not yet suffered in nutrition to an extreme degree. In most cases, no further change takes place, the sounds remaining like those of the fœtus in utero. In typhus fever most commonly, but occasionally in other fevers, a further stage is reached, the first sound losing in tone and finally becoming inaudible. This last condition, however, is extremely rare, as the changes referred to take place on the left or systemic side of the heart chiefly, and the first sound of the right ventricle remains after that of the left has ceased.

Reduplication of the sounds is a common modification. Either sound may be the subject of it. A probable explanation has already been suggested, when the naturally duplicate mechanism of the cardiac sounds was referred to. Such explanation is not, however, entirely satisfactory—does not seem to meet all cases. In mitral stenosis reduplication of the second sound, heard over and below the pulmonary area—often at the apex— is common. When it occurs the usually accompanying accentuation of the pulmonary second sound is less easily recognized. Supposing the two diastolic sounds to be produced on the left and right sides of the heart respectively, it would seem that

the former is the aortic, the latter the pulmonary second sound
(*vide Note,* page 52). Phonetically a double second sound may be
represented by the syllables *ta-ta,* thus with the first sound
Lubb'-ta-ta. In Bright's disease reduplication of the first sound
at the apex and over the ventricles is a very frequent sign.
The ordinary reduplicated first sound may be represented
phonetically by the syllables "turrup;" while the second sound
is represented in the usual way by "dup"—*turrup dup.* A
peculiar form of reduplication is known by the French name
bruit de galop. No verbal description of it would be likely
to succeed in giving the reader an idea of it, but the sound is
very characteristic of a dilated heart, and is well worth careful
clinical study.

In cases of mitral stenosis, the second sound is often
characteristically absent at the apex of the heart. Owing to
this absence of the second sound a variety of presystolic
murmur, followed by an accentuated first sound, is very readily
mistaken by beginners for a systolic murmur followed by the
second sound.

(2) *Extrinsic.*—The heart sounds, as heard over the surface
of the chest, may be modified, not from any change in the
manner of their production, but by interference with their due
conduction to the ear. Hydro-pericardium is an instance of this,
in which the feebleness of the sounds is in marked contrast
with the increased area of dulness. Another significant sign in
such cases is hearing the sounds more distinctly at the upper
part of the sternum, than over the centre of the præcordial
region. Of course the heart's action is interfered with in
these cases and the sounds may be intrinsically altered and
weakened as well. A thick chest-wall necessarily renders the
sounds less audible. The transmission of the heart sounds
beyond the præcordial region which depends upon morbid

alterations in the lungs, is a question properly belonging to the subject of pulmonary diagnosis.

B. *New or Adventitious Sounds.*

I. *Endocardial Murmurs.*— Cardiac murmurs are abnormal, more or less blowing sounds accompanying or replacing the ordinary cardiac sounds, and having a definite relation to physiological action, contraction or expansion, taking place in the chambers of the heart. They vary in loudness and in quality as well as in pitch, but all are prolonged and wanting in the characteristics of a heart sound in the technical acceptation of that term.

For practical purposes murmurs may be divided into two classes—(*a*) Valvular, and (*b*) so-called Hæmic. In the former, which will chiefly engage our attention, there is some structural alteration present at or immediately beyond the cardiac orifice concerned, most commonly in the valves themselves, or there is defect in the muscular complement of the valve apparatus in the case of the auriculo-ventricular valves.

Structural change may give rise to two kinds of murmurs —(*a*) *Obstruction* or direct, and (*β*) *Regurgitation* murmurs. (*a*) Obstruction murmurs may again be divided into two classes —(1) Cases in which the orifice is contracted by local disease, or in which there is some projection from the valves or arterial walls calculated to obstruct the blood-current in greater or less degree. A very loud murmur may be produced by a trifling obstruction of this kind. Such murmurs form the *absolute* obstruction class. (2) Cases of *relative* obstruction, by which term we express the condition present, for instance, in the aorta, when the channel of the vessel is dilated while the orifice retains its normal size, or at any rate is not enlarged proportionately to the channel beyond. Under such circumstances "fluid veins" are formed as the blood current spreads out after its passage through the orifice, and these are the cause of the murmurs audible over the chest-wall.

(β) Regurgitation murmurs may be divided into three classes, according to the mechanism of the incompetence on which they depend. (1) Murmurs produced by incompetence depending upon disease of the valves themselves. (2) Murmurs produced by incompetence of the auriculo-ventricular valves depending on muscle-failure. The auriculo-ventricular orifices have to be prepared by muscle-contraction for the action of the valves, which again are maintained in action by muscle contraction. Moreover, in dilated ventricles the altered relations of the musculi papillares to the curtains have to be taken into account. (3) Murmurs produced at or beyond arterial orifices by incompetence of the valves, the result of dilatation of the orifice consequent upon the dilatation of the adjoining channel of the vessel. Owing to the increased area which the valves have to cover, they, though in themselves healthy, are rendered incompetent.

Endocarditis (generally rheumatic) is the chief cause of deformity of the valves themselves, and this being (except in intra-uterine life) almost confined to the left side of the heart, we have to do with murmurs of the first-class almost alone on that side. A large class of aortic murmurs arise from incompetence of the valves depending on atheromatous changes in the vessel, either as the result of chronic changes in the valves, or of dilatation of the vessel and enlargement of its orifice. Lastly there is the "muscle-failure" class of mitral and tricuspid regurgitation murmurs. The last two classes of murmurs are met with in the degenerative period of life, or under certain special circumstances in earlier life, and bear no relation to rheumatism. The injurious effect exerted upon the

heart-muscle by an adherent pericardium must be borne in
mind. How complicated is the subject of the etiology
of valve lesions may be inferred from the statement of the
fact, that aortic incompetence may depend on half-a-dozen
pathological processes. 1. Rheumatic endocarditis. 2. Ulcera-
tive or septic endocarditis. 3. Chronic sub-inflammatory
changes, accompanied by thickening and subsequent shrinking
of the valves, the result of habitual excessive strain upon
the valves. 4. Atheromatous changes affecting the valves.
5. Atheromatous changes leading to dilatation of the aorta,
and finally of its orifice, so as to render the healthy valves
incompetent. 6. Rupture of a valve (though presumably not
a sound one). The pathological diagnosis of a heart-case
cannot be made by physical signs alone, and must, therefore,
be alluded to as little as possible in a work which deals only
with these.

In the case of every murmur heard over the præcordial region
three points have to be determined :—

1. Rhythm.
2. Position of maximum intensity.
3. Direction of transmission.

1. In every case the rhythm of a murmur should be determined
by keeping the finger on some artery near the heart, as the
carotid or subclavian, while listening.

A murmur may be presystolic (*i.e.*, auricular systolic),
systolic (ventricular systolic), or diastolic (ventricular diastolic).
The following diagram will illustrate the relation of these
different murmurs to the heart sounds :—

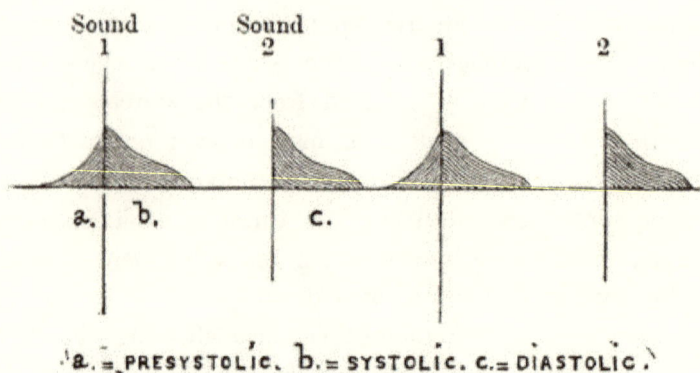

a. = PRESYSTOLIC. b. = SYSTOLIC. c. = DIASTOLIC.

Fig. IX. represents the different rhythms of murmurs as described in the text. With regard to diastolic murmurs of mitral origin, when there is also a presystolic murmur present it is evident that it merely depends upon the duration of these murmurs, whether there be or not a pause between them.

The systolic murmur is represented as running off from the first sound, and when the murmur is due to mitral or tricuspid regurgitation, nature bears out the truth of this representation, if any portion of the first sound remain audible.

Again, in the case of the diastolic murmur of aortic incompetence, it often happens that the murmur is preceded by the second sound, which, of course, is coincident with the commencement of the expansion or the end of the contraction, of the ventricle, while the murmur is coincident with the continuance of the expansion of the ventricle. It is quite common to hear sound and murmur together in the lesser degrees of aortic incompetence. Murmurs of aortic incompetence possess, generally, a *diminuendo* character, that is, they grow fainter from their commencement, as the backward flow may be supposed to become less rapid and forcible, owing to the fuller expansion of the ventricle, and the recovery of the elastic aorta from its distension. Presystolic murmurs, on the other hand, increase in intensity towards the first sound, with which they abruptly close. A tricuspid presystolic murmur is very rare. Constriction of both orifices—mitral and tricuspid—is, however, not very rare, but the degree of constriction of the mitral orifice is usually much greater than that of the tricuspid orifice.

We shall now consider in detail the murmurs which experience has shown to occur in disease.

The Presystolic or Auricular-systolic Murmur is a murmur generated at the auriculo-ventricular orifice, corresponding in time to the physiological action of contraction of the auricle. The ventricle, already near repletion, on receiving the blood more or less forcibly propelled into it by the auricle, at once contracts, so that there is no pause, and the presystolic murmur runs into the first sound, which is always present in these cases. It seems very likely that the flow from the auricle has not ceased at the time when the ventricle contracts and puts an end to the murmur. (By palpation, it will be remembered, the thrill accompanying the presystolic murmur is felt to run up to the apex-beat, with which it abruptly closes.) This murmur increases in intensity up to the first sound, as represented in Fig. IX.; the fact may be explained by the increased power acquired by the auricle as its cavity becomes lessened. However explained, this growing in intensity of the murmur up to the first sound is characteristic. When it possesses all the attributes just described, we receive certain information from the presystolic murmur that mitral or tricuspid stenosis exists. Unfortunately, it is often absent, either temporarily, although, of course, the stenosis remains unchanged, or permanently. The disappearance of the murmur may be of no good omen. It may be due to failure of the contractile vigour of the auricle, but the capricious coming and going of the murmur frequently observed, perhaps scarcely bears out this explanation.

The presystolic murmur, followed by the accentuated first sound of the lesion, may be represented phonetically by the syllable " trrup." It will be remembered that a double first sound is represented by "turrup." It is sometimes difficult to distinguish the presystolic murmur and first sound from a double first sound, but in the latter case there is no *crescendo* character as in the former. A difficulty with regard to the diagnostic value of the presystolic murmur is occasioned by the fact, first pointed out by the late Prof. Austen Flint, that in certain exceptional cases of free aortic reflux, so close a simulation of a presystolic murmur imperfectly developed may be heard as to deceive even a practised ear. The clinical rule is, that in the presence of free aortic regurgitation, and especially when this arises from aortic disease and not from rheumatic endocarditis, great caution must be exercised in the diagnosis of mitral stenosis.

Systolic Murmurs.—In our consideration of the presystolic murmur, matters were rendered comparatively simple, inasmuch as the production of the murmur could take place only at the auriculo-ventricular orifices and have but one mechanism. In the case of systolic murmurs a certain amount of complexity must be encountered, as such murmurs can be generated at either arterial or auriculo-ventricular orifices, and in each case their mechanism is different. Generated at an arterial orifice, the systolic is an obstruction murmur indicating absolute or relative stenosis of the orifice. Arising at an auriculo-ventricular orifice, it is a murmur of regurgitation produced by incompetence of the mitral or tricuspid valves allowing a leakage backwards into the auricle during the systole against the general blood-current. In both cases the murmur coincides in time with the contraction of the ventricles. In quality it varies, being not unfrequently harsh, when due to absolute stenosis and a rough and rigid condition of the aortic valves, and generally soft and blowing when due to incompetence of the auriculo-ventricular valves. In cases of great aortic stenosis, when calcareous valves form a diaphragm across the orifice with only a central slit, this murmur is often peculiarly loud, harsh, and prolonged, while the aortic second sound may be inaudible. Lastly, over a dilated left auricle, it is said that a murmur purely systolic in rhythm is sometimes audible about an inch or an inch and a half to the left of the sternum above the third rib, alleged to be caused by a regurgitated current from the ventricle. This explanation of the murmur heard in the region indicated is very doubtful.

Several points of interest with regard to murmurs of systolic rhythm will call for consideration in the section on the "Transmission of Murmurs." Here we shall only mention the presence, in greater or less integrity, of the first sound in some cases, while in others the murmur entirely replaces the sound. A systolic apex-murmur, not propagated to the back, frequently accompanies mitral stenosis, and is no doubt due to some regurgitation through the deformed orifice. In such cases the first sound is usually present to some extent at the apex, and at the back, indeed behind the mid-axillary line, as a rule takes the place of the murmur. In muscle-failure of the left ventricle, without disease of the valves themselves, an apex murmur is generally not carried to the back. It is often observed in the course of acute rheumatism that a recently developed apex-murmur is accompanied by the first sound at the apex, and is not at first carried to the back, although the first sound may ultimately be lost and the murmur become audible at the back.

Diastolic Murmurs.—Like the systolic, the diastolic is a murmur owning a different mechanism according as it is produced at the aortic or mitral orifices. In the former case it is a murmur of regurgitation, in the latter a murmur of obstruction. As the significance of a systolic murmur is so far exactly the

E

reverse (*i.e.*, produced by blood currents flowing in opposite
directions), according as it is apical or basic in origin, so also
the significance of the diastolic murmur at base and apex is
reversed (*vide* Fig. X.). The diastolic murmur due to aortic

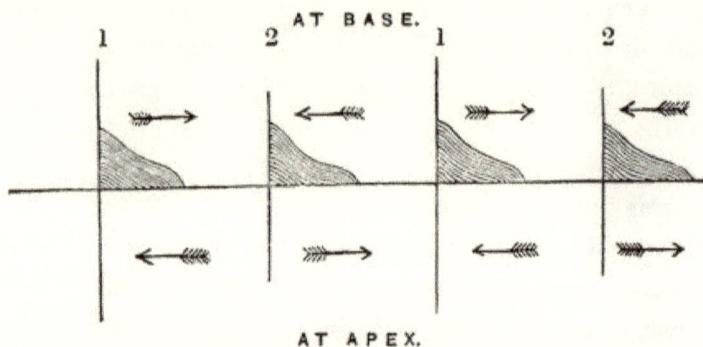

Fig. X.—The arrows represent the direction of the current generating systolic and
diastolic murmurs, according as these are formed at the base or apex of the heart.
Thus a systolic murmur at the base is of obstruction mechanism, at the apex of
regurgitation mechanism. A diastolic murmur at the base, again, is of regurgitation
at the apex of obstruction mechanism. The long vertical lines represent the first
sound, the short the second sound, as in the preceding figure.

incompetence is much the more common, and, as a rule, it is
preceded by a systolic murmur, or at all events, a murmurish
first sound, seldom by a perfectly healthy first sound. Frequently the second sound is audible as well as the murmur,
indicating a less degree of damage to the valve. This occurs
most remarkably where there is dilatation of the first part of
the aorta, and we have a "relative" obstruction murmur,
systolic in time, succeeded by an accentuated second sound,
immediately followed by a blowing murmur. The aortic
diastolic murmur, as a rule, is soft and blowing, but in some
cases it is harsh* and one variety is characterised by being
accompanied by a thrill at the apex of the heart, where the
murmur is specially loud. It is a notorious fact, that under
certain temporary conditions, the aortic diastolic murmur may,

* Aortic diastolic murmurs not very rarely furnish examples of more or less musical
murmurs, occasionally audible at some distance from the patient's body.

for a time, disappear, although the valves are incompetent as ever, as evidenced by the vascular phenomena of the disease. It may be here mentioned that a faint aortic diastolic murmur is the last murmur which the student of auscultation learns to recognise, and even after considerable practice, he will be apt to miss the finer varieties of the murmur.

In rare cases a very short murmur precedes an accentuated second sound over the aorta. Dr. Walshe has represented the combination of sound and murmur by the word PHWE...TT. The sign may be regarded as indicative of a dilated aorta. The murmur is short, and, in comparison with the sound which follows it, insignificant.

The diastolic mitral* murmur differs materially from the pre-systolic murmur already described, not only in rhythm, but in quality. The difference in rhythm will be readily appreciated by reference to Fig. IX. The diastolic murmur does not in-crease in intensity as the presystolic, but on the contrary diminishes. It is commonly harsh, and often accompanied by thrill, as already noted under palpation. This thrill is quite distinct from the presystolic thrill, as already described. Both murmurs are brought about by the same conditions, namely, a narrowed auriculo-ventricular orifice, yet each possesses a mechanism of its own. In the case of the true presystolic murmur we have seen that the contraction of the auricle is the physiological action of the heart associated with the murmur. In the case of the mitral diastolic murmur we have to do with the active (?) expansion of the ventricle *plus* the blood-

* Dr. Balfour states that a diastolic murmur due to mitral stenosis may be audible, and have its maximum intensity in the pulmonary area. This murmur is soft and blowing, unlike the apex true diastolic murmur of mitral stenosis, and is probably produced in the pulmonary artery and infundibulum of the right ventricle as a murmur of high pressure, the pulmonary artery being dilated, and its valves permitting of a certain amount of regurgitation. This murmur is not usually constant, at least when first developed. (*Vid. Med. Chronicle*, Dec., 1888, "The murmur of high pressure in the pulmonary artery.")

·pressure in the lungs. We can readily imagine expansion of the ventricles taking place with greater force at the commencement of the diastole, rendering the current of blood through the auriculo-ventricular orifices more rapid and forcible at the outset, and gradually diminishing in intensity as the cavities are filled. The pressure of blood in the auricle will at the same time diminish. We may expect the murmur to represent the blood current audibly. A presystolic and a diastolic murmur frequently co-exist, when, if there be no pause between them, the diastolic murmur becomes augmented, as the flow of blood through the orifice is reinforced by auricular contraction. If the diastolic murmur exist alone, auricular contraction is in abeyance, as far as murmur-production is concerned.*

2. *The maximum intensity of Murmurs.*—The impossibility of distinguishing the orifice at which a murmur is produced by hearing it most loudly over the anatomical position of the orifice in question, has already been stated and explained. The four artificial areas, which are named respectively after each set of valves, and the reasons for their selection having been already considered, it will suffice here to recapitulate their situations.

The aortic area—over the junction of the second right costal cartilage with the sternum.

The pulmonary area—over the junction of the third left costal cartilage with the sternum.

The mitral area—over the apex-beat.

The tricuspid area—over the lower end of the sternum and portion of right ventricle not covered by lung.

It does not follow that every endocardial murmur must have its maximum intensity over one of these areas. For instance,

---

* When a reduplicated second sound is heard along with the diastolic murmur, the latter portion of the double sound seems heard "through" the murmur as if this followed the former portion.

an aortic regurgitation murmur is not unfrequently audible only over and to the left of the lower part of the sternum, or in the pulmonary area, or at the apex. In fact, a triangle may be sketched out with one side corresponding to the right border of the sternum up to the 2nd cartilage, the other side consisting of a line drawn from the same cartilage to the apex of the heart, and the base of a line drawn from the apex to the end of the sternum, in any part of which triangle an aortic regurgitation murmur may have its maximum intensity or be alone audible ; the whole area must, therefore, be explored in a suspected case. The characters of the sounds or murmurs at each of the named areas must, however, always be carefully investigated and compared with the sounds and murmurs heard elsewhere, while the laws of the propagation of murmurs, to be immediately considered, will materially aid in arriving at a correct diagnosis. Friction rubbing is usually heard best where the heart lies most superficially, and, therefore, in the triangular space which we have described as formed by the indentation in the anterior border of the left lung, but it is, as a rule, developed earliest at the base of the heart. The consideration of the maximum intensity of murmurs is inseparably connected with the subject treated of in the next paragraph—viz., the transmission of murmurs—and will receive in that place further comment. Apart from the isolation of individual murmurs, the position of maximum intensity of the heart-sounds as a whole may be of some importance, as in hydro-pericardium, where the elevation of the area of maximum intensity is significant. All displacements of the heart will alter the position of the maximum intensity of the heart sounds in a corresponding direction. The effect of aneurisms or aneurismal dilatation of the arch of the aorta in intensifying markedly the aortic second sound over their seat affords valuable diagnostic indication. Consolidation of either pulmonary apex may render the cardiac sounds almost as loud there as over the heart itself; but this, and similar phenomena, belong rather to the subject of pulmonary diagnosis, with which we are not at present concerned.

3. *Transmission of Murmurs.*—Much of our success in cardiac diagnosis will depend upon a correct appreciation of this property of murmurs. As has just been stated, the influence of the respiratory organs in conveying the cardiac sounds beyond the heart's own area is great, but here we presuppose the parenchyma of the lungs healthy.* As a general rule, murmurs are carried in the direction of the blood-current producing them.

The normal directions of propagation of the different murmurs are as follows:—

1. Aortic obstruction murmurs—carried upwards in the course of the large vessels.

2. Aortic regurgitation murmurs—carried downwards along and to the left of sternum, often from the level of the third cartilage; sometimes reach the apex and may be little heard elsewhere.

3. Mitral regurgitation murmurs—carried towards the left axilla, and backwards to the vertebral groove.

4. Mitral obstruction murmurs—best heard at the apex, little propagated beyond.

5. A tricuspid murmur—rarely extends much beyond its own area, but inclines to the right.

The amount of first sound accompanying a mitral regurgitation murmur when both are present is liable to vary. A more important point, perhaps, is the change which the murmur is found to undergo on carrying the stethoscope to the left and backwards. Should such a murmur be gradually lost and

---

* The heart sounds, especially the second sound, are usually in health louder under the *right* than the left clavicle. (*Vid. Medical Lectures and Essays,* by Dr. George Johnson, p. 483.) This is no doubt the result of the position of the aorta.

become replaced by a first sound, say at the mid-axillary line and posteriorly to it, a fair presumption may be made that there is either regurgitation through a constricted orifice, or regurgitation from incompetence of healthy valves-curtains as the result of muscle-failure. This rule is not absolute, and the occurrence of a mitral murmur not conducted to the back as a sign of endocarditis in the course of acute rheumatism has already been noted. Moreover, in some cases of mitral stenosis a systolic murmur at the apex is audible at the back. In the presence of rheumatism or chorea, or of a history of either of these, the presumption should always be that any murmur audible over the heart depends on endocarditic changes.

In concluding the subject of endocardial murmurs it is necessary to call attention to a murmur which may be called a "respiratory-systolic" murmur, and which appears not to be produced in the heart at all, but in the superjacent lung by compression. This murmur is usually met with about the apex of the heart, and is therefore liable to be confused with a mitral regurgitation murmur. Its nature will be recognised by the influence exerted upon it by the respiratory act, during some part of which it usually quite disappears. Variation of an apex systolic murmur with respiration does not by any means exclude its being a mitral regurgitation murmur. Regurgitation is possibly freeer during inspiration.

The sounds remaining pure over the area corresponding to the portion of the heart uncovered by lung is an important point in the diagnosis of the "respiratory systolic" murmur.

(b.) *Anæmic Murmurs, &c.*—The consideration of these murmurs occupies a large portion of the space devoted to vascular murmurs; but here passing reference must be made to similar murmurs heard over the præcordial region. Systolic in time, seldom harsh, and never permanently so, they are loudest, and

not uncommonly heard *only* over the pulmonary area. They
may appear, however, at the aortic area, and ultimately at the
mitral and tricuspid areas, although in all probability in the
last two cases, the heart-muscle badly nourished by virtue of
the anæmia, has permitted the valves to become incompetent.*
The chief characteristic of hæmic murmurs is their seat of
maximum intensity over the pulmonary area (where murmurs
the result of structural changes are rare), and their being
commonly accompanied by venous hum in the neck. A great
deal evidently has yet to be learnt with regard to the murmurs
of anæmia not only as to their mechanism, but also with regard
to the blood conditions which induce them. Patients may
suffer from severe degrees of anæmia without developing
murmurs, while others develop marked murmurs when suffer-
ing from only slight degrees of anæmia.

Mention may be made, in this place, of murmurs due to
coagulation of the blood in the chambers of the heart or in the
large vessels. They are generally systolic and loudest at the
base—*i.e.*, arterial obstruction murmurs. Cases, however, have
been recorded where presystolic and aortic diastolic murmurs
have been produced by coagulation : in the latter case, however,
it is doubtful if ulceration of the valves had not caused both
thrombus and incompetence. Common as is the phenomenon
of *ante-mortem* clotting in the heart and large vessels apart from
any local disease, judged by *post-mortem* room experience, it is
very seldom that a murmur is detected during life. No doubt,
in the majority of cases, a murmur, even if produced, would be

---

* The so-called hæmic murmur at the pulmonary area has indeed been supposed
always to be the result of dilatation of the left cavities (mitral regurgitation mechanism),
and really to be formed in an enlarged left auricle and not in the pulmonary artery.
It is said to have its maximum intensity further to the left than a murmur really
generated in the pulmonary artery would have. I am not prepared to assent to this
theory. If the ordinary murmur of anæmia over the pulmonary area is so explained,
the like murmur over the aorta, often almost as loud, remains unexplained.

obscured by the pulmonary râles which in great abundance so often precede death.

II. *Pericardial Friction.*—We have preferred to consider the audible phenomena due to pericardial inflammation apart from endocardial murmurs. The opposed surfaces (visceral and parietal) of the pericardium having lost their normal smoothness and lubrication, the movements of the heart must necessarily be productive of attrition sound, and, inasmuch as the movements of the heart are divisible into systolic and diastolic, so likewise will the rubbing sound produced by these movements be divided. If this friction sound be imperfectly developed, it will probably be deficient in its diastolic portion, the movement of the heart taking place during the diastole being less forcible, and therefore less likely to generate attrition noise. A friction rub may then exist during the systole only, and occur during but a part of this action, while a diastolic friction sound is probably never met with alone. We have already seen that the first and second sounds are not co-extensive in time with the systole and diastole of the heart, contraction and expansion following (and in the case of auricular contraction, preceding) the sounds, so the period of rest enjoyed by the heart must be exceedingly short, unless we consider the act (?) of expansion as rest. From the above considerations we might expect that friction noise would have no definite relation to the heart's sounds, but would extend beyond them, so to speak, and, although divided, be almost continuous. At whatever period there is cardiac movement, at that period also there may be pericardial friction noise. This is always accompanied by a sensation of rubbing, difficult to describe, but readily appreciated at the bedside. Except where the pericardium is covered by lung, friction noise gives to the ear a marked character of being produced superficially. Where, however, the lung overlaps, it may be purely systolic in rhythm, and, from lacking the ordinary character of superficialness, closely simulate an endocardial murmur. From anatomical considerations it might be expected that friction would be most audible over the right ventricle, where it comes into contact with the chest-wall. This accords with the usual situation of the maximum intensity of friction sounds, viz., over the lower half of the sternum and to the left of that region. But Dr. Walshe's statement is well worthy of quotation—"When the entire cardiac surface is the seat of friction murmur, the maximum amount of noise exists, according to some authors, about the nipple ; to others, behind the sternum. *I am satisfied no rule of the kind can be established."* Friction is usually well marked, and occurs early at the base of the heart. One great character of friction sound is its non-transmission in the directions already indicated in the case of endocardial murmurs, or, indeed, much beyond the anatomical limits of the pericardium in any direction. Of course allowance must be made for extreme loudness of the sound. It has been pointed out that attrition sounds do not correspond in time with the cardiac sounds. Another character is their comparative uniformity of intensity from commencement to close. Thus they do not extend up to or run off from the sounds as endocardial murmurs do. In fact they are independent of the cardiac sounds. Occasionally, however, there seem to be "interruptions" of the sounds giving a more or less divided character, it may be to both systolic and diastolic portions. The last, but not least important, peculiarity of friction sounds is their susceptibility to modification by pressure with the stethoscope ; probably this character affords an absolutely diagnostic sign of pericarditis, when a previously single attrition sound is followed by a diastolic sound on making pressure with the stethoscope. It may be well to mention, in conclusion, that the diastolic or second portion of the to-and-fro sound, which constitutes pericardial friction, has been observed to be double, and Dr. George Johnson has observed a presystolic or auricular-systolic friction, giving rise to triple rhythm.* Changes going on in the pericardial surfaces alter from time to time the quality and situation of maximum intensity of attrition sound. Pericardial effusion is in marked contrast with pleuritic effusion in regard to the power of annulling friction sound, very large pericarditic effusions not unfrequently failing to produce this result. Still less likely is effusion to annul friction sound when enlargement of the heart co-exists. The elevation of the apex-beat of the heart in pericarditic effusion has already been noticed, and its influence on the cardiac audible phenomena must not be lost sight of. The vigour of the cardiac contraction exerts much influence upon the production of pericarditic friction noise. In purulent pericarditis there may be no trace of friction sound, for the absence of which the softness of the exudation may partly account, but the chief cause is the greatly depressed vigour of the heart muscle. Of the many varieties of friction sound, extending from a faint graze to a coarse grating sound, no description has been offered, owing to the impossibility of conveying to the mind of those who have not heard for themselves, a correct idea of the sound described.

Sometimes pleuritic friction may be of cardiac rhythm, that is, be produced in the neighbourhood of and by the movements of the heart, but a very little care will obviate any risk of mistaking pleuritic for pericarditic friction.

Vascular Sounds.

1. *Arterial.*—We have already endeavoured to make clear

* "Medical Lectures and Essays." Chap. XXXVI.

the distinction which is drawn between sounds and murmurs in the technical sense of these words. In the case of arterial auscultatory signs, use will still be made of the same terms with the same signification. Many writers speak of the systole and diastole of the heart and of the arteries separately, the systole of the heart of course corresponding to the diastole of the arteries, and the systole of the arteries to the diastole of the heart. This leads almost necessarily to confusion, and here we shall speak of systole and diastole with reference to the heart only.

The arteries commonly examined with the stethoscope, are the carotid, subclavian, and femoral arteries, not because they possess peculiarities, but for purposes of convenience.

The carotid and subclavian vessels being near the heart, it is not surprising that in them we should hear sounds resembling the cardiac sounds, and probably the same transmitted— certainly so as regards the second sound. In the femoral artery an approach to a sound in the form of a thud, systolic in rhythm, is alone audible. This sound is evidently produced in the arterial wall. We have as yet spoken of the auscultatory signs heard over a large artery when pressure is carefully avoided, and the stethoscope laid lightly over the vessel examined. The sound referred to, corresponding with the systole of the heart, whether occurring in the carotid, sub- clavian or femoral, in the great majority of instances *is at once transformed into a murmur on pressure being made* with the stethoscope. The facility with which this change is produced varies much, and in some cases the artery seems to resist the strongest pressure without there being alteration to a murmur. The second sound in the carotids and subclavians remains unchanged, although the first sound be changed into a murmur.

In morbid conditions we may hear either sounds or murmurs

produced in the arteries. We have referred to the systolic
sound audible in health over the femoral artery—a dull thud
rather than a sound. In certain abnormal conditions this thud
comes to resemble a cardiac sound much more closely. This
phenomenon is not pathognomonic of any one condition, unless
it be an abnormally great and sudden change from the state of
relaxation to one of tension, i.e., varying blood pressure. Hence
it is in cases of aortic regurgitation that we meet with the best
examples of this sound. In rare cases of extreme defect of the
aortic valves there is sometimes heard in addition a second
sound. The occurrence of this phenomenon gives us certain
proof that the sounds we hear over arteries can be produced
locally, for in the cases we refer to there exist no cardiac
sounds to be transmitted, if transmission to the femoral artery
were possible.

The potency of an anæmic condition in the production of
murmurs and in the augmentation of those produced by other
causes has been already noticed. Anæmic murmurs are probably
all systolic in rhythm, although it is alleged that a diastolic
anæmic murmur may exist.* Hæmic murmurs, as a rule, are
soft and blowing, although by pressure their character may
be altered. The most important hæmic murmur is the venous,
and while considering it, we shall have to refer again to the
arterial. By means of pressure, we have seen, a systolic murmur can generally be developed over any large artery in health ;
hence it is natural to suppose that from various morbid conditions pressure will occasionally be exerted on arterial trunks, .
and murmurs be thus produced.

Sir Dominic Corrigan long ago showed that murmurs were
produced, not at the point of constriction, but on the distal side

---

* I have on one or two occasions examined cases which *seemed* to support the
allegation. The subject of diastolic murmur in anæmia is referred to in Dr. Stephen
Mackenzie's *Lettsomian Lectures, 1891.*

beyond.   We have already dwelt upon the importance of
"relative obstruction" in treating of the murmurs of the heart.

On pressure being made over the large arteries in free aortic
regurgitation, a double murmur, systolic and diastolic, is found
to be the rule, but the diastolic portion is usually very much
shorter and less loud, while a certain degree of pressure is
required to develop it.   No doubt, as heard over the carotid
and subclavian arteries, these murmurs may be transmitted
from the aortic orifice and its immediate neighbourhood.   Direct
murmurs formed at the aortic orifice are, in accordance with the
common rule, carried onwards in the course of the circulation
with much greater facility than regurgtation murmurs.   The
double murmur heard in the femoral must, however, be generated
locally.   By pressure on the vessel we produce obstruction to
two currents, the one flowing onwards, much the more powerful,
the other flowing backwards and feeble, corresponding with the
regurgitant flow backwards through the valves.   In each case,
" fluid veins " are generated, audible as murmurs.

Auscultation of the Veins.

Venous.—In anæmic subjects over the internal jugular veins at
the root of the neck, and especially on the right side, the venous
current becomes soniferous.   *Murmur* is produced, and the murmur differs from all the murmurs which we have considered in
being *continuous.*   The quality and intensity of the sound varies
much, not only in different cases but in the same case, by applica-
tion of varying degrees of pressure, and according to the rapidity
of the bloodcurrent returning to the heart.   Among the agencies
tending to accelerate the blood-current in the jugulars may be
mentioned inspiration, the diastole of the heart, and the upright
posture.   Inasmuch as these act independently of one another the
intensity of the murmur is constantly liable to slight variations,

irrespective of any change in the amount of pressure exerted. In quality the murmur may be soft and blowing, humming or musical, or it may consist of an almost roaring noise. Apart from the intrinsic variation in intensity, of which we have mentioned some of the causes, the murmur seems regularly intensified in many cases during the systole of the heart. When this is the case it will usually be found that we are dealing not with a venous murmur alone, but also with the arterial murmur already described. To an unpractised ear this combination bears some resemblance to a double murmur. The arterial murmur, drowning the venous for the time, represents the systolic portion, while the venous murmur, inaudible from the loudness of the arterial murmur during the systole, alone represents the diastolic portion of the double murmur.

In the above sentences we have described the venous murmur as occurring at the root of the neck in an anæmic subject. Naturally we have taken as our example the phenomenon where best developed ; but other veins, as the subclavian, femoral, superior longitudinal sinus, &c., manifest similar murmur. Moreover, chlorosis or anæmia is not necessary for a slight development of the sign. Nay, it is met with, usually feebly marked however, in cases in which little or no departure from health can be detected. On the other hand all forms of anæmia do not induce the murmurs usually associated with the condition.

3. *Auscultation of Aneurismal Sacs.*—Either sounds or murmurs, or sounds and murmurs together, may be audible over aneurismal sacs. So many variations and different combinations are found to exist that Dr. Walshe enumerated no fewer than ten varieties which he had himself heard, adding that "the list probably might be increased from the experience of others." With regard to murmurs or sounds audible over

aneurisms of the arch of the aorta, the questions, of course, will always arise : to what extent are the sounds or murmurs generated at the aortic orifice ? to what extent in the sac itself ? To commence with the simplest type, we may hear over the pulsating area exceedingly loud sounds, the second invariably possessing more or less of the characteristic feature we have called accentuation. That the aortic valves should be closed with exaggerated force is only what we might expect. The existence of two sounds, resembling (apparently being) the heart sounds, and at least of as great intensity, away from the region of the heart itself, is a sign strongly presumptive of aneurism ; and, when heard over an area pulsating with at least as great force as the heart, the presumption of aneurism rises almost to certainty. In considering the propagation of murmurs, the influence of the conducting power of the lungs has already been mentioned. Here it is only referred to, lest error may arise from abnormal loudness of the heart sounds over a consolidated apex. The distinctness of the first sound, however, in aneurisms varies much, there being frequently present only a dull thud. Sometimes there is only a systolic jog or push, which is felt rather than heard. The second sound also may be deficient in distinctness, though still conveying the impression of accentuation. In aneurism of the arch it is the systolic sound which commonly fails. The next combination we have to refer to is a systolic murmur, followed by the accentuated second sound so often mentioned. Thirdly, there may be a double murmur. In the case of aneurisms of the arch of the aorta, of course in the great majority if not in all, when double murmur exists, the aortic valves are incompetent, and it has been supposed that both murmurs are transmitted from the heart. It is probably by no means always so, and it is not unlikely that murmurs may be produced about the mouth of saccular aneurisms. The systolic is usually the louder and harsher of the two murmurs, although the diastolic may be the more prolonged. Another feature often presented by the double murmur heard over aneurismal sacs is its continuity— the systolic and diastolic portions being practically continuous. These remarks sufficiently indicate the auscultatory signs which may be expected to be found over aneurisms of the arch, and aneurisms in this situation can alone be considered in a description of the " Physical Signs of Cardiac Disease."

APPENDIX.

THE PULSE.

For purposes of convenience we usually place our finger on the radial artery at the wrist to ascertain the rate of the heart beats, and from this universal custom we have come to speak of *the pulse*, meaning the pulsation felt in the radial artery. Each beat of the pulse corresponds with a cardiac systole, but they are not quite synchronous. As we pass from the centre to the periphery, a perceptible interval becomes apparent between the cardiac and vascular beats. In certain morbid conditions this interval becomes more evident, and in aortic reflux the interval is very pronounced, so much so, as in some cases to render the pulse almost synchronous with the succeeding ventricular contraction, instead of with its own. It is a good precaution to feel both radial pulses in investigating vascular disease. Absence or diminution of the radial pulse on one side is an important sign of aneurism. In many cases, however, it depends on irregular distribution of the arteries merely, and sometimes upon obstruction due to endarteritis. Both subclavian and carotid pulses have been known to be obliterated by chronic endarteritis at the origin of the vessels. In every case absence or diminution of a radial pulse is a sign calling for further investigation.

The pulse may be considered under the following heads :—

Derangements or Variations of the Pulse.

A.—With reference to the Pulse, considered as a series of Beats or Impulses.

B.—With reference to the Pulse, considered as a Single Beat.

C.—With reference to the condition of the Arterial Walls.

A. includes irregularity in respect to either

(*a.*) Frequency, } or both together.
(*b.*) Equality,

B. includes the following variations :—

1. Large and small.
2. Full and empty.
3. Compressible and incompressible.
4. Quick and slow.

5. Dicrotism, &c.
6. Thrills.
7. Corrigan's, or the water-hammer pulse.

A.

(*a.*) *Frequency of the Pulse Beat.*—Pulse-rate, though of comparatively minor importance in cardiac diagnosis, cannot be altogether passed over without comment. In the majority of cases of valvular disease of the heart the pulse is rendered more frequent. This effect is commonly observed in mitral disease, while one form of aortic disease is characterised by the opposite tendency (aortic obstruction). Fatty degeneration of the heart sometimes reduces the pulse much below the normal frequency. Of an infrequent pulse each individual beat may be quick and of short duration, or, on the other hand, slow and laboured. The point is of some clinical importance, and should always be noted in a report. The normal standard is from 60 to 70 per minute, but many individuals present variations from this standard, some having a pulse as infrequent as 30, others as frequent as 80 and more. In the condition known as "tachy-cardia" the pulse becomes extremely frequent—200 per minute for instance. When it is remembered that the pulse-rate forms a feature in all the specific fevers, and is not without diagnostic and prognostic value in most other diseases, the multifarious bearings of its study will be apparent; but here we have to do with pulse-rate only in its relation to the diagnosis of diseases of the heart.

(b). *Equality.*—In health one beat is as another. In disease this may no longer be the case, and some beats may be full and strong, while others are hardly perceptible, and some heart-beats may even be imperceptible in the pulse. The arrangement of the dissimilar beats varies indefinitely.

Irregularity of the Pulse.—This common feature of the pulse in cardiac disease may be considered as composed of two elements, one having relation to the pulse *frequency*, the other to the pulse *equality*. The well-known occurrence of inter-mission may be regarded as a defect in frequency. It is, in a large majority of cases, a pause in the rhythmic contractions of the heart; but it may be due to inequality carried to excess—*i.e.*, the ventricular contraction being so feeble as not to be represented in the radial artery. Distinct pauses may be followed by a series of rapid beats, or there may be no pause yet the pulse vary in frequency and force at different moments. A strong beat commonly follows a pause, (*vide Fig.* 15) and then a series of weaker beats may immediately ensue. Irregularity of the pulse is a feature of cardiac disease, easily recognised and of much importance, although it is not necessarily the expression of valve disease. When, indeed, it is associated with the last, it probably depends more upon the state of nutrition and innervation of the heart substance than upon any special form of valve defect, although it is usually regarded as characteristic of mitral disease. Irregularity is by no means uncommon in the absence of all organic diseases of the heart, as in dyspepsia, functional derangement of the liver, the specific fevers, &c. Sometimes a succession of irregular beats recurs in a cycle, forming a sort of order in disorder. A not uncommonly observed phenomenon under the physiological action of digitalis, especially in cases of mitral stenosis, is an apparently very slow pulse, which is found on further investigation to be accompanied by a normal frequency of the cardiac contractions; when the latter are more closely studied in relation with the pulse, they are found to occur in pairs, one of each pair alone producing a pulse beat in the radial artery (*vide Fig.* 16). Although most common in cases of mitral stenosis, this peculiarity of the pulse is met with in other diseases. It has been observed, for instance, in a case of aortic regurgitation under the influence of digitalis. In mitral stenosis it may occur independently of the action of the digitalis.

The forms under which irregularity may be met with are exceedingly numerous, and no advantage would accrue from an attempt to enumerate them.

The *"Pulsus Paradoxus."*—This pulse is characterised by becoming very feeble during inspiration. It has been stated to be characteristic of certain cases of "chronic mediastinitis" associated with adherent pericardium.

B.

The characters of the individual beat of the pulse have of late, perhaps, ceased to be the object of interest which they were to the older physicians. The invention of the sphygmograph is the chief cause of this, in the same way that the invention of the clinical thermometer may be charged with the decline of attention directed to-day to pulse frequency in the specific fevers and febrile diseases generally. The comparative disuse of these simple methods of diagnosis and prognosis has undoubtedly been carried too far.

The Individual Pulse Beat.—Under this head we enumerate a few of the leading varieties of *pulses.*

1. *Small and Large.*—The meaning to be attached to these terms is apparent, and clinically the distinctive character of each is easily recognised. In mitral disease the pulse is usually small. In dilatation of the left ventricle with hypertrophy it is, as a rule, large. There can be little doubt that when the walls of the ventricle have become degenerated and the cavity has become rounded in shape, the systole is often habitually incomplete. It must be remembered that, normally, the supra papillary space of the left ventricle contains blood at the end of the systole. The smallness or largeness of the pulse does not, however, depend solely upon the condition of the left ventricle. For instance, general exposure to cold will render a normal pulse small by contracting the muscular coat of the vessels, while warmth tends to produce an opposite effect.

2. *Full and empty.*—By an empty pulse we mean one that gives us, on palpation, the sensation of being but partially filled, or, in other words, that the vessel is capable of holding more than is supplied to it. The full pulse, on the other hand, is the direct counterpart of this, for in it the blood seems supplied to repletion.

3. *Compressible and Incompressible.*—At the bedside it will be found that pulses vary much as regards the amount of pressure necessary to cause their obliteration. The pulse which resists pressure is "incompressible," the pulse readily obliterated "compressible." The degree of compressibility is an important sign, and is always worthy of attention. In the days of bleeding much attention used to be directed to this.

character, and the incompressible pulse, perhaps wisely, formed an indication for the lancet. The incompressible pulse is *par excellence* the pulse of "high tension." In a good example of a high tension pulse the radial artery is felt like a solid cord, and when the finger is placed lightly on it no pulsation may be perceptible in it. Pressure with the finger not only elicits the pulsation but shows the high degree to which the vessel may be compressed without the pulsation being obliterated. The expression "full between the beats" is often applied to the high tension pulse. High tension is associated with peripheral resistance, but requires a vigorous left ventricle for its maintenance.

4. *Quick and Slow.*—These characters are most manifest in the infrequent pulse, of which we may have two varieties. One is characterised by a rapid impulse of short duration, the other by a prolonged impulse. The conditions of peripheral resistance and of the contraction of the left ventricle no doubt are the chief factors in the production of these varieties. When the pulse is interfered with by the interposition of an aneurism between it and the heart the pulse is prolonged; there is delay of the beat but not of its beginning "but of the period of maximum pressure." (Fig. 13.)

5. *Dicrotism.*—Sometimes there is a double beat corresponding to a single cardiac contraction, the second beat being very much less perceptible. This is the dicrotic pulse, and one usually believed to be associated with low tension. Drs. Roy and Adami deny this. (*Vid.* Practitioner, 1890, p. 134.) It is commonly met with in enteric fever and other febrile states. (Figs. 3, 4, 5.)

6. The pulse in some morbid states is accompanied by thrill. This happens occasionally in cases of aortic regurgitation, with dilated ventricle and unfilled arteries.

7. *Corrigan's Pulse, or the Pulse of Aortic Reflux.*—This pulse possesses a very special character, and bears the name of the physician who first described it—Sir Dominic Corrigan. It is known also under the names of the " water-hammer pulse," " the pulse of unfilled arteries," &c. As already mentioned, this pulse possesses an extreme degree of visibleness. It is characterised by a sudden filling of the artery followed immediately by an equally sudden emptying, and this peculiarity becomes exaggerated on raising the arm to a right angle with the recumbent trunk. In cases of aortic regurgitation, especially the variety depending on endocarditis, the pulse in the peripheral arteries is frequently delayed, it may be to such a degree that the pulse becomes almost synchronous, not with its own systole, but with the one succeeding.

C.

Endo-arteritis.—Having ascertained the characters of the pulse proper, there remains to be investigated still another point, namely, the state of the arterial walls. This is readily ascertained by obliterating the pulse by firm pressure, and rolling the forcibly-emptied vessel against the bone, when, if the coats be thickened and atheromatous, we are able at once to recognise the condition. In some cases the vessel becomes almost moniliform, from irregular thickening and calcareous deposit in its coats.

For further particulars regarding the pulse the reader is referred to the series of sphygmograms which follows.

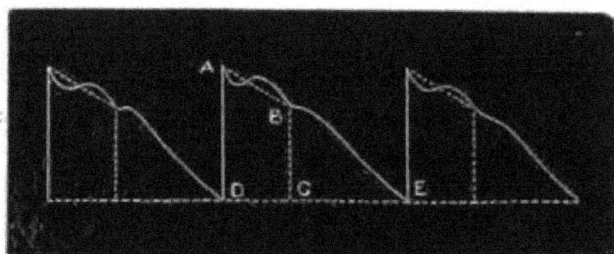

FIG. 1.—(The late Dr. Mahomed.) Shows a method of gauging tension. The tracing will serve to illustrate the principal features of a pulse curve. There is the "upstroke," then a depression forming with the upper portion of the "upstroke" the "percussion wave." An elevation, the "tidal wave," succeeds. This wave terminates with the "aortic notch" (B) which indicates the end of systole and the beginning of diastole. The elevation immediately following this notch is the "dicrotic wave." A line drawn through the lower extremities of the series of "upstrokes" is called the "respiratory line." The tracing represents a pulse of considerable tension, and it will be noticed that the "tidal wave" rises above the dotted line drawn from the summit of the "upstroke" to the "aortic notch." The "aortic notch," again, is at a considerable height above the "respiratory line." The old and best known nomenclature is retained here. Drs. Roy and Adami, in their paper already referred to, propose to name the "percussion wave" the "papillary wave" and the "tidal wave" the "outflow-remainder wave." Their observations, further, seem to throw some doubt upon the suitableness of the terms "high" and "low tension," in common use, as applied to tracings presenting the characteristic features, which are illustrated in the following tracings.

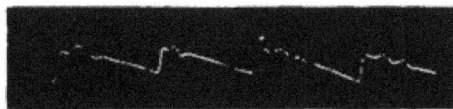

FIG. 2.—(Mahomed.) Showing high tension. In this case there was no albumen in the urine.

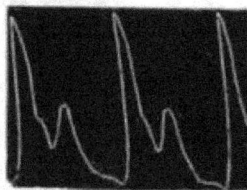

FIG. 3.—(Mahomed.) Low tension and dicrotic pulse. The aortic notch approaches the respiratory line.

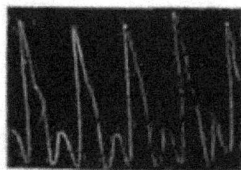

FIG. 4.—(Mahomed.) Fully dicrotic pulse. The aortic notch reaches the respiratory line.

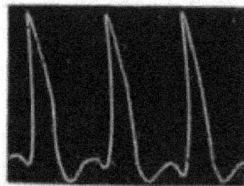

FIG. 5.—(Mahomed.) Hyperdicrotic pulse. The aortic notch falls below the respiratory line.

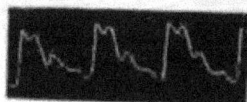

FIG. 6.—(Mahomed.) Tracing showing pronounced tidal and dicrotic waves.

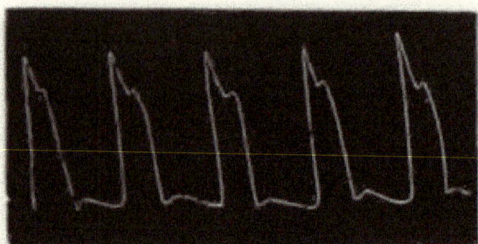

FIG. 7.—(Mahomed.) Tracing of pulse in aortic regurgitation due to endocarditic disease.

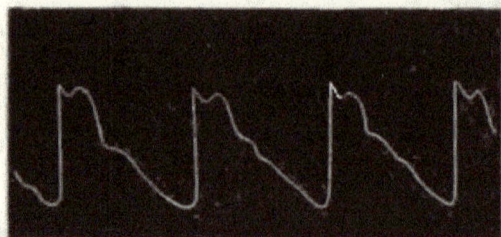

FIG. 8.—(Mahomed.) Pulse in case of aortic regurgitation, due to aortitis deformans.

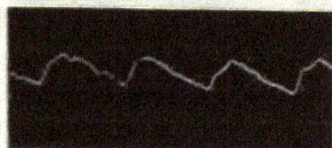

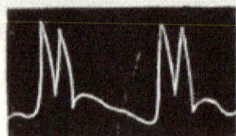

Fig. 9.—(Mahomed.) Tracing of pulse in aortic obstruction.

FIG. 10.—Tracing of pulse in aortic obstruction showing another type. (From Reynolds' "System of Medicine," Vol. IV., p. 639.)

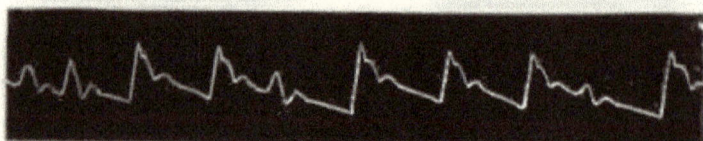

FIG. 11.—Pulse in mitral stenosis. The tracing shows a well-marked tidal wave and the occasional occurrence of abortive beats.

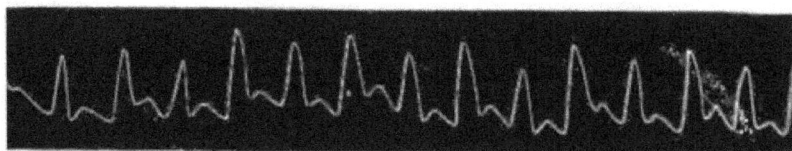

Fig. 12.—(Mahomed.) Pulse of cardiac dilatation, with mitral regurgitation. There is no tidal wave. The upstokes vary in length.

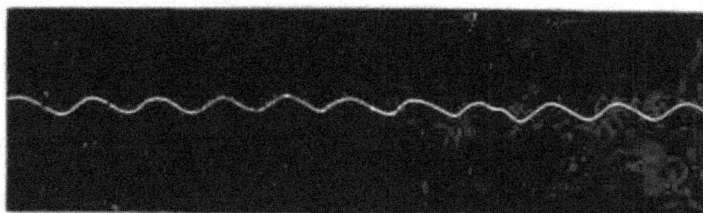

Fig. 13.

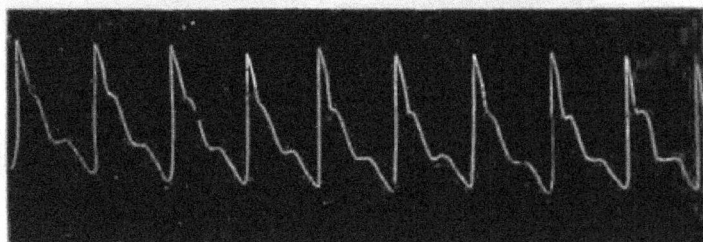

Fig. 14.

Figs. 13 and 14 (Mahomed) represent the right and left radial pulses respectively in a case of aneurism of the right axiliary artery. The former is extremely aneurismal in character.

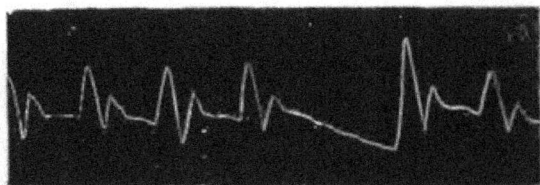

Fig. 15.—Tacing of pulse showing a "missed" beat with succeeding unusually full beat. The tracing was taken from a case of great dilatation of the left ventricle. The systole was no doubt habitually imperfect. As a consequence of the missed beat and the accumulation of blood in the chamber, the left ventricle seems to have been roused to a supreme effort.

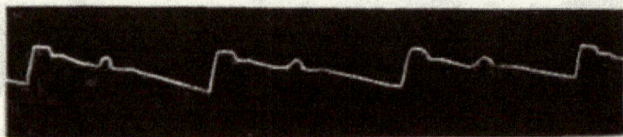

FIG. 16.—Bigeminal pulse. If the abortive heart-beat is imperceptible in the pulse the latter seems only to be unusually slow.

FIG. 17.—(Mahomed.)   Pulse of atheromatous arteries.

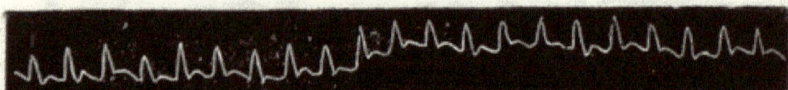

26th Oct., 1887.   FIG. 18.

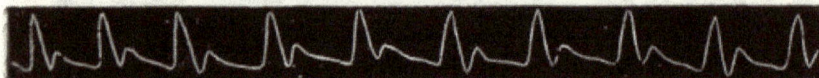

25th Oct.   FIG. 19.

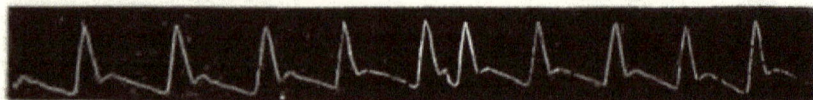

1st Nov.   FIG. 20.

19th Nov. FIG. 23

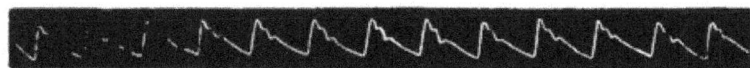

6th Dec. FIG. 24.

FIGS. 18—24 represent pulse tracings taken in a case of "heart-failure" occurring during the course of chronic Bright's disease. The patient made a good recovery from his "heart-failure." The tracings shows a progressive increase of pulse-tension. (Case reported in *Medical Chronicle*, Jan., 1888.)

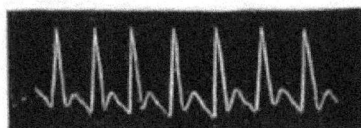

FIG. 25.

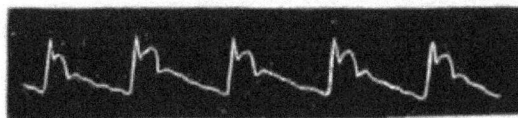

FIG. 26.

FIGS. 25 and 26.—Pulse tracings from a case of "heart-failure."
FIG. 26 represents the pulse after recovery. (Case reported in *Lancet*, Aug. 14th, 1886.)

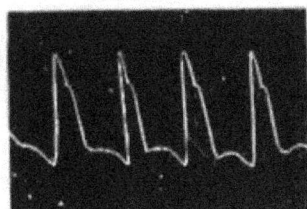

FIG. 27.

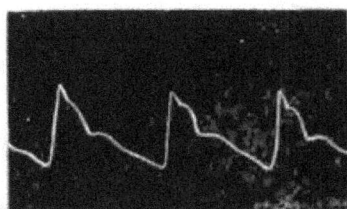

FIG. 28.

FIGS. 27 and 28.—Pulse tracings from a case of "cardiac-failure."
FIG. 28 represents the pulse after recovery. (Case reported in *Lancet*, Sept. 10th, 1887.)

T. SOWLER AND CO.,
PRINTERS, CANNON STREET,
MANCHESTER.

Roger Bacon. The Philosophy of Science in the Middle Ages. By R. Adamson, M.A., Professor of Logic and Mental and Moral Philosophy in the Owens College, Victoria University. 1s.

Greek Exercises for Beginners. Translated from the Greek Grammar of Prof. George Curtius. By Edwin B. England, M.A., Lecturer in Greek and Latin in the Owens College, Victoria University. 1s.

Diseases of the Bones: Their Pathology, Diagnosis, and Treatment. By Thomas Jones, F.R.C.S., B.S., Lecturer on Practical Surgery in the Owens College, Victoria University. 12s. 6d.

The Present Aspect of the Antiseptic Question. By Edward Lund, F.R.C.S., Prof. of Surgery in the Owens College, Victoria University. 2s.

The Frog : An Introduction to Anatomy, Histology, and Embryology, with a Chapter on Development added. By A. Milnes Marshall, M.D., D.Sc., M.A., F.R.S., Professor of Zoology and Comparative Anatomy in the Owens College, Victoria University. Third Edition, Revised and Illustrated. 4s.

Owens College : Studies from the Biological Laboratories. Vols. I. and II. Published by the Council of the College and Edited by Prof. Milnes Marshall. 10s. each.

Engineering Syllabus of the Lectures at the Owens College, together with a series of examples relating to the various subjects included in the course. By Osborne Reynolds, M.A., F.R.S., Prof. of Engineering in the Owens College, Victoria University. Arranged by Mr. J. B. Millar, Assistant Lecturer in Engineering. Second Edition. 3s.

Description of the Chemical Laboratories at the Owens College, from the plans of Alfred Waterhouse, R.A. By Sir H. E. Roscoe, F.R.S., Professor of Chemistry in the Owens College, Victoria University. With lithographed copies of the original plans and elevations. 5s.

On Technical Education. Inaugural Address as President of the Medical Society of Manchester, 1889. By James Ross, M.D., LL.D., Professor of Medicine in the Owens College, Victoria University. 1s.

Studies from the Physiological Laboratory of the Owens College. Vol. I. Published by the Council of the College and Edited by Prof. William Stirling, M.D., Sc.D. 10s.

Histological Notes for the use of Medical Students. By W. Horscraft Waters, M.A., Lecturer on Histology in the Owens College, Victoria University. 2s. 6d.

www.ingramcontent.com/pod-product-compliance
Lightning Source LLC
Chambersburg PA
CBHW020311090426
42735CB00009B/1305